ISSN 0021-1915

16. JAHRGANG · SEITEN 313—344 · BERLIN OKTOBER 1980

IPRAX 16 16 (10) 313—344 (1980)

32036

10

isotopenpraxis

EVP 9,— M

AKADEMIE-VERLAG

An unsere Autoren

Die Zeitschrift veröffentlicht Beiträge zu folgenden Themen: Kontrolle und Analyse radioaktiver und stabiler Isotope, Anwendung radioaktiver und stabiler Isotope in allen Gebieten der Forschung und Technik, der Industrie, Landwirtschaft und Medizin, Herstellung radioaktiver und stabiler Isotope sowie von markierten Verbindungen, kernelektronische Geräte für die Isotopenanwendung (Dickenmesser, Dichtemesser u. ä.) Detektoren für isotopentechnische Arbeiten und den Strahlenschutz in Isotopenlaboratorien, Meßtechnik und Strahlenschutz, Dosimetrie in Isotopenlaboratorien, kerntechnische Geräte für die Isotopenanwendung (Betron, Bestrahlungsanlagen u. ä.), Strahlenchemie, Radiochemie, ökonomische und perspektivische Fragen, Fragen der Weiterbildung.

1. Aufnahme von Manuskripten

Zur Veröffentlichung angenommen werden nur druckreife, noch nicht anderweitig publizierte Beiträge in deutscher, englischer oder russischer Sprache zum Themenbereich der Zeitschrift und zwar

Originalarbeiten
Kurze Originalmitteilungen
Aus der Praxis
Zusammenfassende Berichte
Tagungsberichte
Informationen.

Sind der Beitrag oder Teile daraus bereits veröffentlicht oder ist eine anderweitige Veröffentlichung vorgesehen, so muß dies der Redaktion mit genauer Angabe der Zweitveröffentlichung mitgeteilt werden. Mit der Annahme des Manuskriptes behält sich der Verlag das Recht der Vervielfältigung, Verbreitung und Übersetzung innerhalb der gesetzlichen Schutzfrist vor. Auszugsweiser Nachdruck oder photomechanische u. a. Wiedergabe sind an eine schriftliche Genehmigung des Verlages gebunden. Manuskripte, Zeichnungen und Photos von veröffentlichten Arbeiten werden nur auf den besonderen und bis zum Zeitpunkt der Veröffentlichung mitgeteilten Wunsch des Autors zurückgesandt.
Bei der unverlangten Einreichung von Manuskripten wird empfohlen, eine Befürwortung eines der Herausgeber oder Redaktionsausschußmitglieder beizufügen.
Manuskripte sind einzureichen bei der Redaktion, DDR-1080 Berlin, Leipziger Str. 3–4.

2. Form der Manuskripte

Die Autoren werden gebeten:
2 Manuskripte, eines davon in Original-Maschinenschrift, einseitig und mit 2zeiligem Abstand einzureichen; gut leserliche Photokopien sind zulässig; auf dem 1. Blatt (unter Weglassung der Bezeichnung)
den Titel der Arbeit
Vorname-Initialen und Namen sämtlicher Autoren
gegebenenfalls die Institution, in der sie tätig sind
eine kurze Zusammenfassung der Arbeit von höchstens 15 Zeilen
Keywords in alphabetischer Reihenfolge und
als Fußnote die Anschrift der Institution anzugeben;
wenn möglich auf gesonderten Blättern Namen, Titel der Arbeit und Zusammenfassung in den beiden anderen obengenannten Sprachen 2fach beizufügen; Literaturhinweise (fortlaufend numeriert), Fußnoten, Tabellen, Abbildungen (in etwa 2facher Veröffentlichungsgröße), Abbildungsunterschriften (getrennt von den Abbildungen, mit Zeichenerklärungen, Quellenangaben, falls erforderlich, und gegebenenfalls mit einem Vermerk für besondere Wünsche bezüglich der Abbildungsgröße), private und Dienstanschrift der Verfasser mit akademischen Titeln und ausgeschriebenen Vornamen auf besonderen Blättern ans Ende des Manuskriptes zu stellen;

Wünsche für den Kolumnentitel (Seitenüberschrift) bekanntzugeben;
Absätze durch Einzüge von mindestens 3 Anschlägen kenntlich zu machen;
die Maßeinheiten des SI-Systems ("Systéme International") zu verwenden;
Korrekturen im Manuskript unmittelbar an der zu korrigierenden Stelle anzubringen;
Formeln deutlich lesbar und möglichst groß einzutragen;
Farbabbildungen und nachträgliche Änderungen oder Ergänzungen in Abbildungen zu vermeiden;
hervorzuhebende Textstellen zu unterstreichen, und zwar halbfetten Satz mit einer geraden schwarzen Linie, kursiven Satz (Schrägschrift) mit einer blauen Wellenlinie (alle Eigennamen). Sperrungen gestrichelt;
besondere Buchstaben durch farbige Unterstreichung zu kennzeichnen, und zwar Formelzeichen in

Griechisch-Kursiv ($\alpha, \beta, \ldots$) rot,
Kursiv (A, B, ...) blau
halbfetter Kursiv ($\boldsymbol{A}, \boldsymbol{B}, \ldots$) 2 Linien blau (Vektoren),
halbfetter Grotesk (**A**, **B**, ...) 2 Linien violett (Tensoren);

bei Literaturangaben auf Vollständigkeit zu achten. In nachstehender Reihenfolge müssen angegeben sein:

a) bei *Zeitschriften*: Vorname-Initialen und Namen sämtlicher Autoren Titel der Zeitschrift (nach „Periodica Chimica" oder möglichst wenig gekürzt), Bandnummer (unterstrichen), Jahreszahl (in Klammern), Heftnummer (nur bei Zeitschriften, die jedes Heft mit S. 1 beginnen) sowie die Seitenzahl, auf der die zitierte Arbeit beginnt.

b) bei *Büchern*: Initialen der Vornamen und Name des Autors (Herausgeber werden durch den Zusatz „ed." gekennzeichnet), voll ausgeschriebener Titel des Buches in der Originalsprache, evtl. Bandnummer, voll ausgeschriebener Name des Verlages, Nummer der Auflage (außer der ersten), Erscheinungsort(e), Erscheinungsjahr (hier nicht in Klammern setzen!) und evtl. Kapitelnummer, Seitenzahl u. ä.

c) bei *Sammelbänden, Tagungsberichten*: Es soll zusätzlich der Titel des zitierten Beitrags angegeben werden. Die Seitenzahl muß unbedingt angegeben werden. Aus dem Zitat muß hervorgehen, ob es sich um einen Tagungssammelband oder um einen noch nicht veröffentlichten Vortrag handelt. Zwischen Tagungsort und Erscheinungsort des entsprechenden Tagungsberichts ist zu unterscheiden.

3. Korrekturen

Der Autor erhält zwei Umbruchabzüge, von denen einer möglichst schnell korrigiert an die Redaktion zurückgereicht werden soll. Auf Wunsch des Autors kann die Korrektur auch von der Redakion gelesen werden.

Durch nachträgliche Manuskriptänderungen verursachte Korrekturkosten gehen, sofern sie 10% der Satzkosten übersteigen zu Lasten des Autors.

Der Autor erhält 50 Sonderdrucke kostenlos. Weitere Sonderdrucke können bis zur Rückgabe der Korrekturabzüge auf Kosten des Autors bei der Redaktion bestellt werden.

Bezugsmöglichkeiten

Bestellungen sind zu richten

- **in der DDR** an den Postzeitungsvertrieb; an eine Buchhandlung oder an den AKADEMIE-VERLAG, DDR-1080 Berlin, Leipziger Straße 3–4
- **im sozialistischen Ausland** an eine Buchhandlung für fremdsprachige Literatur oder an den zuständigen Postzeitungsvertrieb
- **in der BRD und Westberlin** an eine Buchhandlung oder an die Auslieferungsstelle KUNST UND WISSEN, Erich Bieber, 7000 Stuttgart 1, Wilhelmstraße 4–6
- **in Österreich** an den Globus-Buchvertrieb, 1201 Wien, Höchstädtplatz 3
- **in den übrigen westeuropäischen Ländern** an eine Buchhandlung oder an die Auslieferungsstelle KUNST UND WISSEN, Eirch Bieber GmbH, CH-8008 Zürich/Schweiz, Dufourstraße 51
- **im übrige n Ausland** an den Internationalen Buch- und Zeitschriftenhandel; den Buchexport, Volkseigener Außenhandelsbetrieb der Deutschen Demokratischen Republik, DDR-7010 Leipzig, Postfach 160, oder an den AKADEMIE-VERLAG, DDR-1080 Berlin, Leipziger Straße 3–4

Zeitschrift „Isotopenpraxis"

Herausgeber: Im Auftrag des Akademie-Verlages herausgegeben von Prof. Dr. Justus Mühlenpfordt und Prof. Dr. Carl Friedrich Weiss
Verlag: Akademie-Verlag, DDR-1080 Berlin, Leipziger Straße 3–4; Fernruf 2236221 oder 2236229; Telex-Nr. 114420;
Bank: Staatsbank der DDR, Berlin, Kto.-Nr.: 6836-26-20712
Chefredakteur: Dr. Klaus Uhlig, Berlin
Anschrift der Redaktion (Fernruf: 6351177) der Zeitschrift Isotopenpraxis: Akademie-Verlag, DDR-1080 Berlin, Leipziger Str. 3–4
Veröffentlicht unter der Lizenznummer 1527 des Presseamtes beim Vorsitzenden des Ministerrates der Deutschen Demokratischen Republik
Gesamtherstellung: VEB Druckerei „Thomas Müntzer", DDR-5820 Bad Langensalza
Erscheinungsweise: Die Zeitschrift „Isotopenpraxis" erscheint monatlich. Die 12 Hefte eines Jahrganges bilden einen Band. Bezugspreis je Band 180,– M zuzüglich Versandspesen (Preis für die DDR 108,– M), Preis je Heft 15,– (Preis für die DDR 9,– M)
Bestellnummer dieses Heftes: 1079/16/10

AN (EDV) 12831

ISOTOPENPRAXIS ISSN 0021-1915

Band 16 (1980) Heft 10, S. 313—344

Redaktion und Verlag: Akademie-Verlag, Redaktion Isotopenpraxis, 1080 Berlin, Leipziger Straße 3—4

Inhaltsverzeichnis

ISOTOPENPRAXIS

Band **16** (1980) Heft 10, S. 313—344

Redaktion und Verlag: Akademie-Verlag, Redaktion Isotopenpraxis, 1080 Berlin, Leipziger Straße 3—4

Contents

H. R. Schuette

The Application of the Tracer Technique in Agricultural and Related Sciences (*in English*)

Isotopenpraxis **16** (1980) No. 10, p. 313 to 320
5 Fig., 1 Tab., 125 Ref.
This article reviews the principles of tracer technique and its application in agricultural and related sciences involving counting radioactivity, synthesis of radioactive labelled compounds and its use in solving analytical, metabolic and biosynthetic problems.

Progress Reports

W. Stolz

Lyoluminescence Dosimetry (LLD)

Isotopenpraxis **16** (1980) No. 10, p. 321 to 325
5 Fig., 1 Tab., 24 Ref.

Contents:

1. Introduction
2. Mechanism of lyoluminescence (Alkali halogenides; Organic compounds)
3. Lyoluminescence measurement
4. Dosimeter substances and their properties (Dosimeter substances; Dosage-measuring range; Energy dependence of gamma-radiation; Neutron sensitivity; Fading; Use of lyoluminescence dosimeters)

Bibliography

Articles

R. Kunze, O. Hecker

Determination of the Mean Expended Energy $\overline{W}_L$ for the Formation of an Ion Pair in Air Related to ^{60}Co-Gamma-Radiation

Isotopenpraxis **16** (1980) No. 10, p. 325 to 327
— Fig., 3 Tab., 9 Ref.

On the basis of an ionometric and a calorimetric method of determination of the absorbed dose in graphite, the $\overline{W}_L$ value related to ^{60}Co-gamma-radiation has been determined in ASMW. The following values have been received in dry air:

$$(\overline{W}_{L/e}) = (33.87 \pm 0.34)\ J/C$$

The errors of the method are discussed. The result is compared to recent published values.

T. Szabová, E. Mišúrová

Changes of Nucleic Acids in Blood of Rats after Single X-Ray Irradiation (*in Russian*)

Isotopenpraxis **16** (1980) No. 10, p. 327 to 330
2 Fig., 1 Tab., 18 Ref.

Quantitative changes of nucleic acids and in vivo incorporation of ^{3}H-thymidine and ^{14}C-orotic acid (3 h) in blood of rats were studied during 7 days after single x-ray irradiation with 154,8 mC/kg (600 R).
A steep decrease of DNA concentration up to the 12th hour after irradiation accompanied by the decrease of bound tritium activity in blood were found, the specific DNA activity, however, did not change.
During the following days DNA concentration remained on the decreased level. ^{3}H-thymidine incorporation was increasing stepwisely, and on the 7th day after irradiation the specific DNA activity reached control values over 5-fold approximately. Changes in RNA were of little evidence and mostly nonsignificant.

Содержание

Х. Р. Шютте

Применение метода меченых атомов в сельскохозяйственной и сродных отраслях науки (*на англ. языке*)

Isotopenpraxis **16** (1980) вып. 10, стр. 313 до 320
5 рис., 1 табл., 125 лит.

Данная работа дает обзор о принципах методов меченых атомов и их применении в сельскохозяйственной и сродных науках. Рассматриваются проблемы измерения радиоактивности, синтеза меченых соединений и его применения в целях решения аналитических, метаболических и биосинтетических проблем.

Аналитические обзоры

В. Штольц

Лиолюминесцентная дозиметрия (ЛЛД)

Isotopenpraxis **16** (1980) вып. 10, стр. 321 до 325
5 рис., 1 Табл., 24 лит.

Оглавление:

1. Введение
2. Механизм лиолюминесценции (Щелочные галогениды; Органические соединения)
3. Лиолюминесцентное измерение
4. Дозиметрические вещества и их свойства (Дозиметрические вещества; Диапазон измерения дозиметра; Зависимость гамма-излучения от энергии; Чувствительность к нейтронам; фединг; Применение лиолюминесцентных дозиметров)

Литература

Статьи

Р. Кунце, О. Хеккер

Определение средней потерянной энергии $\overline{W}_L$ по образованию пары ионов в воздухе применительно к ^{60}Co-гамма-излучению

Isotopenpraxis **16** (1980) вып. 10, стр. 325 до 327
— рис., 3 табл., 9 лит.

На основе ионометрического и калориметрического метода определения поглощенной дозы в графите в АСМВ выявлено значение $\overline{W}_L$ применительно к ^{60}Co-гамма-излучению.
По отношению к сухому воздуху были получены следующие величины:

$$(\overline{W}_{L/e}) = (33,87 \pm 0,34)\ J/C.$$

Обсуждаются погрешности метода. Результат сопоставляется более новым опубликованным величинам.

Т. Сабова, Е. Мишурова

Изменения нуклеиновых кислот в крови крыс после однократного рентгеновского облучения (*на русск. языке*)

Isotopenpraxis **16** (1980) вып. 10, стр. 327 до 330
2 рис., 1 табл., 18 лит.

Исследовались количественные изменения нуклеиновых кислот и инкорпорация ^{3}H-тимидина и ^{14}C-оротовой кислоты »ин виво« (3 часа) в крови крыс в течение 7 суток после однократного облучения рентгеновскими лучами в дозе 154,8 *мС/кг* (600 *Р*). Наблюдалось резкое понижение концентрации ДНК до 12-ого часа после облучения, сопровождаемое понижением активности связанного трития в крови, однако удельная активность ДНК существенно не изменялась. В течение последующих дней концентрация ДНК оставалась на сниженном уровне. Инкорпорация ^{3}H-тимидина постепенно возрастала так, что на 7-ой день после облучения удельная активность ДНК была почти в 5 раз выше контрольных величин. Изменения РНК были мало выразительными и в большинстве случаев несигнификантными.

The Application of the Tracer Technique in Agricultural and Related Sciences*) **)

H. R. Schütte

(Academy of Sciences of the GDR, Institute of Plant Biochemistry, Halle/S.)[1])

This article reviews the principles of tracer technique and its application in agricultural and related sciences involving counting radioactivity, synthesis of radioactive labelled compunds and its use in solving analytical, metabolic and biosynthetic problems.

Keywords

agriculture; biosynthesis; carbon 14; labelled compounds; metabolism; radiometric analysis; reviews; tracer techniques; tritium

1. Introduction

Isotopes, especially radionuclides, are an useful tool for labelling chemical compounds in order to look for their fate in chemical or biochemical reactions [1, 2], because the physical and chemical properties of radionuclides are considered to be exactly the same as those of nonradioactive nuclides within a deviation due to the isotope effects. This tracer method is elaborated by v. *Hevesy* and *Paneth*, who have carried out their classical work with natural lead isotopes in 1913 at the Radium Institute in Vienna [3]. The discovery of this tracer method is one of the most important results of natural science. Without this method the big success of the last decades in different topics especially of the applied chemical topics like biochemistry, medicine, biology and agricultural science were not thinkable. Besides the use of radionuclides as tracer atoms they were also used as a source of radiation for chemical and biological reactions. This topics will not be discussed here. We will only discuss the tracer method in which radiation dependent reactions are undesirable side reactions.

Very important for the broad development of the tracer method is the production of convenient isotopes. Highlights on this pathway were the discovery of nuclear reactions by *Rutherford* [4], of the artificial radioactivity by *Curie* and *Joliot* [5], of the use of neutrons for nuclear reactions by *Fermi* [6], and the discovery of the nuclear fission of uranium by *Hahn* and *Straßmann* [7, 8]. The construction of nuclear reactors with a very high neutron flow was necessary for the production of the needed radionuclides. Thus ^{14}C is produced by irradiation of nitrogen-14 with neutrons and a following emission of protons [9]. In a similar manner tritium is yielded from lithium by irradiation with neutrons and a following emission of α-particles [10].

$$^{14}_{7}N(n,p)^{14}_{6}C; \quad ^{6}_{3}Li(n, \alpha)^{3}_{1}H$$

Till now the familiar unit of radioactivity was the curie (Ci), which is equal to $3{,}7 \cdot 10^{10}$ disintegrations per second. But this may disappear with the introduction of the SI unit of the becquerel [Bq], which is equal to 1 disintegration per second. These units are very important because the radioactivity is bought in mCi or Bq or dependent units. More important for calculating the radioactivity in the mass is the so-called specific radioactivity which is the ratio of radioactivity to mass of molecules containing isotopes of the specified radionuclide, usually expressed in curies or millicuries, now in bequerel per millimole. The specific activity must be of a value large enough to assure a high sensitivity required from experimental aspects and small enough to make self-radiolysis of negligible importance under the experimental conditions and to carry out the experiment safely.

Most of the commercially available samples of labelled compounds are of an order of magnitude of 10^{-3} to 10^{2} Ci/mmol for their specific activity being sufficiently large for most of the radiotracer experiments. Sometimes there is a need for a higher specific activity in order to achieve even higher sensitivity, e.g. protein-binding assays for determination of extremely small amounts (say, 10^{-12} mol) of antigens, haptens, hormones, vitamins, and drugs.

The maximum specific activity is given in the so-called carrier-free state. Such an upper limit for ^{14}C- or ^{11}C-labelled compounds, for example, is 62,4 and $9{,}18 \cdot 10^{9}$ mCi/m atom of carbon and for ^{32}P is $9{,}12 \cdot 10^{3}$ mCi/m atom. Apparently, the shorter the half-life, the more conveniently the radiotracer experiment can be designed, although too short-lived radionuclide is rather difficult to handle and often impossible to convert into a useful chemical form. From a practical standpoint, the carrier-free radionuclides, are not always rountinely available.

The development of separation methods for stable isotopes was the basis for having these isotopes in a highly enriched form, especially for nitrogen and oxygen because of these elements no radionuclides are available with convenient half lives [11]. Both kinds of isotopes, the stable and radioactive isotopes have advantages and disadvantages. The stable isotopes are not dependent on the experimental time. The absence of radiation allows their application to living organisms without danger. The disadvantage of the stable isotopes are the not very convenient methods of measurement. The samples have to be destroyed and the possible dilution factors are not very high. In contrast to this the radionuclides have the radiation which is harmful to the health. Otherwise the measurement of the

*) Lecture held during a FAO/IAEA Study Tour on Nuclear Techniques and Induced Mutations in Plant Genetics and Breeding, GDR, 16–24 June, 1979

**) HerrnProf. Dr. Dr. h. c. mult. K. Mothes anläßlich der 80. Wiederkehr seines Geburtstages in Verehrung gewidmet

[1]) Address: DDR-4010 Halle/S., Weinbergweg

Tab. 1. Important isotopes for biochemistry

Element	Stable isotope	Radioactive isotope	Half-life	Radiation
Hydrogen	$^{2}H = D$	—	—	—
Hydrogen	—	$^{3}H = T$	10,46 a	β, very weak
Carbon	—	^{11}C	20 min	β
Carbon	^{13}C	—	—	—
Carbon	—	^{14}C	5730 a	β, weak
Nitrogen	^{15}N	—	—	—
Oxygen	^{18}O	—	—	—
Phosphorus	—	^{32}P	14,28 d	β
Phosphorus	—	^{33}P	25,5 d	β, very weak
Sulfur	—	^{35}S	87,1 d	β, weak
Chlorine	—	^{36}Cl	$3 \cdot 10^{5}$ a	β
Bromine	—	^{82}Br	35,5 h	β
Iodine	—	^{123}I	13,3 h	γ, weak
Iodine	—	^{125}I	60,2 d	γ, weak
Iodine	—	^{131}I	8,05 d	β

radioactive radiation is rather easy; thus very low amounts of these isotopes can be estimated. The dilution factors are bigger than for stable isotopes. With a normal apparatus for detection ^{14}C labelled compounds with a mole mass of 100 till 200 and a specific radioactivity of 2 mCi/mmol can be estimated in a range of still 5×10^{-9} g.

Tab. 1 shows the most important isotopes in biochemistry. We see that there are very big differences in half life [12] and also in the energy of the radiation. The choice between the different isotopes will depend on half life, cost, ease of labelling, sensitivity needed etc. Radionuclides with a short half life have some importance in medical application [13].

2. Radioisotope counting techniques

^{14}C and tritium which are now the most widely used isotopes in the biological sciences, emitted very weak β-rays. Therefore special methods are necessary for the measurement. The radioactive radiation can be detected generally on the basis of their action. The most important action is the excitation or ionization of other atoms or molecules [14]. In the case of the excitation one electron in an atom is raised to a higher energy level. If this electron is going back to the basis situation the liberated energy is emitted as an electromagnetic radiation and can be measured. During the ionization one electron is excited in such a way that it is emitted from the atom and it gives a current impulse in an electric field. If the voltage is increased the ionized gas particles are accelerated to a point where they can cause secondary ionization by collision with nonionized gas molecules giving a gas amplification factor between 10^{5} and 10^{8}.

The most important radiation detector working according the ionization method is the *Geiger-Müller*-counter. With special windows in the counter it can be used also for carbon-14. But it is not convenient for detecting the tritium radiation. Similar effects are the basis for the so-called flow counters which are working mostly with methane as a counting gas. They are very important especially in the thinlayer scanner in which an open detector with methane as counting gas can be droven about the chromatograms.

In the modern scintillation counting the scintillation light is measured via a photomultiplyer. We have to differentiate between the crystal counting and the liquid scintillation counting. The increase in the use of crystal (sodium iodide) or gamma counting arises from the emergence of radioimmunoassay as the predominant technique in the quantitation of substances of biological significance. γ-photons produce no direct ionization by collision along their path as they are devoid of mass, but are absorbed by one of the following three mechanism, the photoelectric effect, the Compton effect, and by pair production. The γ-photons are best absorbed by materials of high electron density. Thallium-activated sodium iodide due to the high *Z* number of the iodine atom has been found to be superior in absorbing γ-photons efficiently, and emitting light photons proportional in intensity to the energy of the absorbed γ-photons.

The liquid scintillation counting is very convenient for the detection of the weak radiation preferably of carbon-14 and tritium [15–19]. In this method the radioactive source together with the scintillator are solved in a convenient solvent, e.g. toluene or dioxan. This intimate mixing of the radioactive sample with the scintillator results in high detection efficiency for weak β-emitters. The energy transfer process β-decay → solvent → scintillator → photomultiplier, which transforms radiant energy originating from radioactive decay to electrical energy by the photomultiplier is a near-linear process.

This ability to relate the radiant energy absorbed to the size of the electrical pulse emerging from the photomultiplier is the basis for the possibilitiy to differentiate between different radionuclides, to determine two isotopes simultaneously in the same sample in the modern liquid scintillation spectrometers. This is very important for some problems in which double labelled compounds are used.

The scintillator for liquid scintillation counting should exhibit a high efficiency of light production and the emitted light should be at a wavelength that corresponds to the area of maximum sensitivity of the photomultiplier. One of the most useful scintillators is 2,5-diphenyloxazol (PPO, POP) together with secondary scintillators. The quantitation of the amount of radioactivity present in a sample in absolute units (dpm) depends on the knowledge of the sample counting efficiency which is usually determined using corresponding standards and which depends preferably on the different quenching of the sample, that means a certain absorption of scintillator light photons in more or less coloured solutions. Therefore coloured solutions, as pikrates, urine or body and plant fluids have to be decoloured before counting.

For materials that cannot be directly incorporated into a toluene solution, some techniques for sample preparation are used [20, 21]. Emulsion counting is used for aqueous samples. The most widely used emulsion system is based on a mixture of toluene and Triton X-100. But several commerical products as Instagel and others are developed, which can accomodate up to 80% water. Solubilization procedures are used for substances such as proteins, which are normally insoluble in water or the base organic solvents used in counting solutions. The different solubilizing agents, as Hyamine and others, are all strong quaternary ammonium bases. In the combination techniques the tritium and/or ^{14}C compounds were oxidized, now in semiautomatic combustion apparatus, to H_2O and CO_2 which were counted after absorption in corresponding solutions. CO_2 is absorbed preferably in an organic amine.

^{32}P can be measured also by Cerenkov counting which is currently useful for energetic β-emitters [22–25]. *Cerenkov* light is generated when a charged particle travels through a medium faster than the speed of light through that medium.

Another method for detection of radioactivity is the autoradiography, in which labelled compounds can be visualized by blackening of a photographic emulsion [26, 27]. By this method it is very easy to estimate the distribution of agrochemicals in plants or of radioactivity of chromatograms for example for testing the purity of synthetic compounds. The chromatogram is put on a X-ray film and after a corresponding exponation time and developing we can see black spots indicating radioactive compounds. A modification of this procedure which is especially important for tritium with its weak β-radiation is scintillation autoradiography or fluorography, which involves interaction of the β-particle with an organic scintillator by mixing or spraying a scintillator solution onto the chromatogram [28, 29]. In the microautoradiography for determing precisely the location of the labelled molecules in cells this technique has been combined with both light and electron microscopy. In these cases tritium is favored because the low energy and penetrating power of its β-particles permits the greatest resolution. The great advance in pure and applied cytology are due largely to the application of these techniques [30–34].

3. The preparation of radioactive labelled compounds

In order to get radioactive labelled compounds the commercial companies offer a lot of different substances. But especially in the field of drugs for pharmacology or in agricultural sciences sometimes it is necressary to synthesize special compounds [1,35–40]. In dependence on the experiment to be made we have the choice of different nuclides, of the position of labelling and specific activity. In the common choice between tritium and ^{14}C, cost, ease of labelling, and sensitivity all favor tritium. Tritium is also technically better for autoradiography at the cellular and subcellular levels. But if it is used as an accessory tracer for carbon it is necessary to think carefully about the stability of carbon-hydrogen bonds involved. Labelling for studies of the mechanism of chemical or biological reactions should be made at nonlabile positions, otherwise correction must be made for intra- and/or intermolecular transfer of the specified atom due to isotopic exchange.

We can distinguish preferably three methods for the synthesis of radioactive labelled compounds, the classical synthetic, the biosynthetic and the radiochemical methods. The first is the most important. This method has the advantage that according to the pathway specific labelled (e.g. in Fig. 1 only in the carboxylic group) and according to the applied specific radioactivity also high labelled compounds are available. The two most important factors in practice are the small scale of working and the rather high cost of the isotopes.

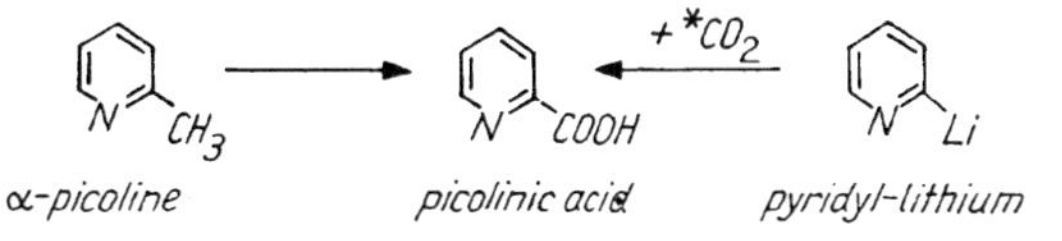

Fig. 1. Radiochemical synthesis of picolinic acid

All ^{14}C-syntheses are starting from carbon dioxid, but a wide range of ^{14}C labelled synthetic intermediates may now be purchases commercially. For the synthesis of radioactive labelled compounds preferably of ^{14}C-compounds according to classical methods sometimes special pathways are to be going. For example for the synthesis of picolinic acid a normal chemist would oxydize the commercially available α-picoline. The radiochemist has to go according to a pathway in which the radioactivity could be introduced. Because the radioactivity is rather expensive it should be introduced at a very late step. Thus the picolinic acid could be synthesized in radioactive labelled form by synthesis of an organic metallic compound like pyridyllithium which was treated with radioactive CO_2, according Fig. 1. In this way a picolinic acid is synthesized which is specific labelled in the carboxylic group [41]. This type of reaction is useful also for other carboxylic acids.

For the synthesis of tritium labelled compounds we know preferably the hydrogenation of double bounds, the decarboxylation of carboxylic acids, the catalytical hydrogenation of halogen compounds and the catalytical exchange method. Especially in the first method there is always some risk that exchange or hydrogen migration will occur at the same time. The hydrogenation of halogen compounds is useful also for complicated compounds if it is possible to introduce in an easy way chlorine or bromine which can be substituted by tritium gas (Fig. 2). With tritium gas of high specific activity thus the corresponding tritium labelled compound is available with high specific activity. In the decarboxylation reaction mentioned the exchange reaction of the carboxylic acid in tritium labelled water is used (Fig. 3) [42]. But the yielded tritium labelled compounds with an activity in the ionic proton of the carboxylic group is not useful for experiments because after solution in water it is reexchanged against hydrogen. But this unstable tritium in the carboxylic group will be stable after decarboxylation if from the carboxylic hydrogen a stable C–T-bond is achieved. With tritium labelled compounds we have sometimes the trouble with the so-called

$$R-X+T_2 \xrightarrow{Pd\ or\ Pt} RT+TX$$

Fig. 2. Catalytic dehalogenation by tritium

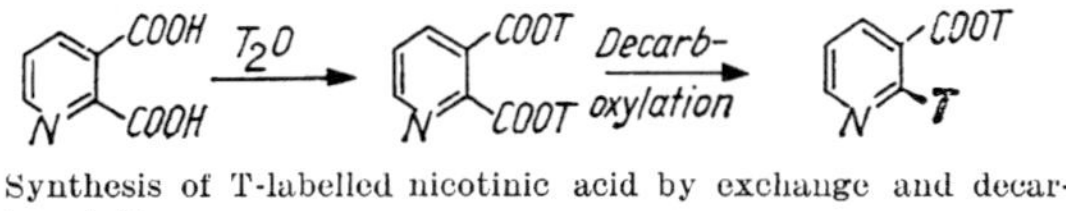

Fig. 3. Synthesis of T-labelled nicotinic acid by exchange and decarboxylation

unstable tritium e.g. in COOH, NH_2, OH but also in CH_2 groups in neighbourhood of carbonylic groups which can become enol groups by keto-enol-tautemerism. This unstable tritium has to be removed by evaporation sometimes with water or other OH containing substances.

The catalytic exchange method is often the easiest and quickest method of tritium labelling. The unlabelled organic substance is heated, with for example, tritiated water, acetic acid, benzene or tritium gas [43] and a catalyst such as palladium or platinum. Whereas this method results in unspecific labelling a similar method with phenols and basic or acidic catalysts is yielding rather specific labelled compounds, in which preferably the ortho and para position are exchanged [44, 45].

For the chemical synthesis also enzymatical methods are used giving stereochemical specific labelled compounds, for example the prochiralic stereochemical specific labelled mevalonic acids [46]. Double labelled compounds, e.g. with ^{14}C and T, can be obtained by going the synthetic pathway with the two isotopes, or more easier by mixing the corresponding single labelled substances.

For the biosynthetic methods we use the synthetic capacities of different organisms, preferably plants and microorganisms. A convenient precursor is given to the organism who is synthesizing the interesting substance. For example radioactive CO_2 is given to a photosynthesizing plant in a closed chamber; then this plant makes the photosynthetic sugars in radioactive labelled form and we have to extract, to separate and to purify the wished compound. For these methods the experiment should be begun if the wished compound is present only in a very low concentration. For example for the biosynthesis of sugars by photosynthesis we let deplete the leaves before the experiment by living in the dark that the containing sugars are respirated. Another very important synthesis with biosynthetic methods is the formation of radioactive labelled amino acid yielding the natural stereoisomers [47]. In this case microorganisms or algae were used and we can begin the experiments only with some cells which are propagated in a radioactive cultural medium. The difficulty of the biosynthetic methods are in any case the extraction, separation and purification of the wished compound. In contrast to the classical synthetic methods in the biosynthetic methods starting with CO_2 we get so-called uniform labelled compounds which have the disadvantage that we cannot differentiate between the single C-atoms but they have the advantage that we can get very high specific activity by summarizing the activities of the single C-atoms. Use of suitable organic substrates gives much more possibilities of biosynthetic labelling.

Of the radiochemical methods only the tritium labelling according *Wilzbach* should be mentioned [37, 40, 48, 49]. In this case in a closed evacuated apparatus the finely divided organic compound is exposed to high active tritium gas. In this apparatus a hydrogen exchange occurs between the tritium gas and the hydrogen atoms of the compound, but the reaction is complicated by side reactions and, above all, by extensive radiation decomposition. As a consequence, a very complex mixture of labelled compounds results, and special care is necessary in purification. Nevertheless this method has a certain importance because it is relatively simple to try, although it is seldom that of choice.

It is essential that the tracer compound used has a purity value which is suitable for the experiments being undertaken and for which it has been prepared. Because of self-decomposition [50, 51] also by self-irradiation this quality is not maintained on storage of the compound. Therefore it is necessary for investigators to check the quality of the radiochemical immediately prior to the experiments.

4. Application of the tracer technique

Which are the uses of radioactive labelled compounds especially in agricultural sciences? The different uses are understandable if it is imagined that the radioactive isotopes have the same chemical properties as the normal elements, but we can estimate the radioactivity in a very simple way. But also the differences between the isotopes, which are expressed in the kinetic isotope effects, has become a widely employed and powerful tool in furthering our understanding of reaction kinetics both in the gas phase and in solution [52–54].

Radionuclides have been used extensively in environmental studies [55–60], for example for water velocity measurements [61], measurement of reaeration of water [62] in studies related to the migration of living things for understanding of movement, reproduction rate, and other ecological factors related to terestrial and aquatic animals as well as insects. In each case a certain quantity of a suitable radionuclide is placed on or in an animal in such a way that the animal is not injured and its behaviour is not significantly altered.

A very big field of use is the analysis with radioactive labelled substances (radiometry). Over the last three decades the use of analytical radioisotope techniques has undergone a continous growth [63]. With the knowledge of the specific activity of a compound it is possible to calculate the weight of an unknown amount of this compound by measurement the radioactivity. This method is very important for studying transport and distribution behaviour in agricultural science, in medicine or in physicochemistry. A radioactive labelled compound or tracer particle is introduced into the system under study. The accumulation into or secretion from the system of the radioactivity is used as a measure of the process or transport of the inactive compound.

Thus the method is applied in the analysis of pesticides and fertilizers in persistence and penetration experiments in soils and in uptake, translocation and excretion experiments in plants and animals [64–69]. By using radiotracer compounds, it is possible to study the fate of agricultural chemicals at very low concentrations. Thus the fate of the radiolabelled herbicide bentazon has been studied in rats by use of whole-body autoradiography techniques [70]. A similar technique was used to study the body distribution of ^{14}C after rats were dosed with radiolabelled α-sulfbenzyl penicillin [71]. The in vivo measurement of the distribution of radioactivity in the body is now used routinely in medical diagnosis [72, 73]. Graphic records of the accumulation of radioactivity within select regions of the body will often produce images that may be used to assess the location, size, shape, or function of organs or special tissues. Thyroid imaging with ^{131}I was the first routine use of scanning. The test takes advantage of the fact that the thyroidal cell has a highly efficient mechanism for the accumulation of iodine. Injection of Dextran or plastic micropheres labelled with radioisotopes into intravascular compartment has for many years been used in circulation studies on test animals. Clinical interest has been directed in particular to the selective retention of different labelled compounds in tumors [74, 75].

Very often these radiometric analytical methods were used in enzymology [76–80], based upon the conversion of a labelled substrate to a labelled product. The rate of the enzymatic reaction is then determined, using the specific activity of the substrate, after separating substrat and product and measuring the radioactivity of either. Therefore the two major requirements in setting up a radiometric enzyme assay which has sometimes many advantages over conventional assays, are the availability of labelled substrate of known specific activity, and a simple and rapid method of effecting a quantitative separation of labelled substrate and product. This separation can be achieved preferably by precipitation, by methods involving the release or uptake of activity in a volatile form as tritiated water or $^{14}CO_2$ by solvent extraction methods or by chromatographic and absorption methods.

A special type of the radiometric analysis is the isotope derivative method. In this procedure a labelled reagent of known specific activity is used to form a labelled derivative of the compound being estimated. The amount of compound may be determined simply by measuring the activity of the product and dividing this by the specific activity of the reagent or by use of a calibration curve. The reaction to the derivative may be achieved also by enzymatical methods. An example is the radiometric determination of L-canaline, a non proteinogenic amino acid from Canavalia ensiformis. In this case 2-^{14}C-acetone is employed as reagent which is volatile while its reaction product with L-canaline, a *Schiff'* base is not. The residue radioactivity is proportional to the L-canaline content [81]. Another example is the detection of small weights of protein with dansylchloride [123, 124] or the estimation of catechol amines by methyltransfer from ^{14}C- or ^{3}H-labelled S-adenosylmethionin using either catechol-O-methyltransferase or phenylethanolamin-N-methyltransferase [82, 83]. The limit sensitivity of these assays will depend on the specific activity of the labelled reagent. Liquid scintillation counting permit the measurement of 0,1 nCi and, with specifically labelled ^{14}C compounds available at $\sim 50\ \mu$Ci/μmol and ^{3}H-compounds available ~ 20 mCi/μmol this gives a detection of 2 pmol with ^{14}C and 0,005 pmol with ^{3}H.

In the double-isotope derivative method a trace of a high specific activity labelled form of the compound under assay is added and then a derivative is formed with a reagent labelled with a second radionuclide such that the activity of each nuclide can be measured in the presence of the other. Complete recovery of product is not essential and the unknown amount of material present in the original sample may readily be calculated after isolation of a radiochemically pure sample of the doubly labelled derivative and measurement of the radioactivity of the two nuclides in it. The derivation can be achieved either chemically or enzymatically. An example of this technique is an assay of methionine [84]. The procedures depends upon the addition of tracer quantities of L-(methyl-^{3}H)-methionine and its conversion to (^{3}H, ^{14}C)-S-adenosyl-L-methionine in the presence of (^{14}C)-ATP and of purified Escherichia coli ATP: L-methionine S-adenosyltransferase.

The double isotope derivative method leads to another special example of the radiometric analysis, the isotope dilution analysis for the quantitative estimation of substances [85]. In this method a trace radioactive labelled form of the compound under assay is added to the unknown mixture and afterwards the compound being estimated is separated from the mixture. It is not necessary, and this is an advantage, to separate the compound quantitatively, it is only necessary to separate an aliquote part but this in a pure form. From the mass of the added labelled substance (X_2), its specific activity (A_2) and the specific activity of the diluted sample (A_1) after separation of the mixture we can calculate the unknown mass of the substance in the mixture according

$$X_1 = \left(\frac{A_2}{A_1} - 1\right) \cdot X_2$$

The isotope dilution method is applied for the quantitation of components in biological fluids, like red cells and plasma in blood, whole-blood volume or total body-water determination using tritiated water.

A special case of the dilution method is the reversed isotope dilution analysis which consists in adding quantities of a known nonisotopic pure compound to impure isotopic material of unknown structure. Recrystallization or some other form of purification is carried out repetitively until a constant specific activity is obtained. The presence of isotope in the isolated material under these conditions indicate an identity between labelled and unlabelled material. This method is used for example for the identification of only small quantities of impure materials of the metabolites in research of the metabolism of phytoeffectors and other drugs in agricultural or medical science. In most cases the phytoeffectors will be metabilosed in living systems. Since the applied amounts are very low in concentration it is difficult to make a classical analysis of the metabolism products. Therefore the probable transformation product in not radioactive form is given to the mixture to have some amounts for handling. If the applied phytoeffector is radioactive labelled then also the transformation products must be radioactive. If the added compound is identical with the transformation product, then after separation and purification also the added compound must be radioactive by mixture with the identical transformation product.

Special methods of the isotope dilution techniques are the different competitive protein-binding assays [86–92, 125]. Of the various types of these assays, radioimmunoassays are clearly the most versatile since antibodies can be developed against almost any substance desired, often in extremely high titers and with very high association constants. The radioimmunoassay based upon the principle of immunological reactions, that an antigen (Ag) and antibody (Ab) give an antigen-antibodycomplex (Ag · Ab). Is the unlabelled antigen (Ag) substituted by a labelled antigen (Ag^+), which is chemical identical or similar, but is not to differentiate immunological of the unlabelled one, it is obtained the corresponding complex ($Ag^+ \cdot Ab$). If the amounts of Ab and Ag^+ are kept constant, the amount of $Ag^+ \cdot Ab$ will be inversely proportional to the amount of Ag added. This relationship may be shown by plotting the percent of bound Ag^+ against the amount of nonradioactive Ag added. If known amounts of a standard are used to obtained such a curve, then the amount of an unknown may be read from the curve.

The radioimmunoassay combines the unique specificity of the antigen-antibody reaction with the sensitivity of a trace determination of radioactivity. For many low and high molecular weight compounds it provides the most sensitive analytical method available, with detection limits between $10^{-12} - 10^{-15}$ mol of antigen. Many of the developed radioimmunoassays are commercially available in analytical sets, so called kits.

In other competitive protein-binding assays the antibody can be substituted by other proteins like transport proteins (transin assay) different receptors (receptor assay) or enzymes (enzymatic assay).

The first radioimmunoassays were developed for insulin [93] and other peptides, which are antigenic. But a large number of radioimmunoassays now in use measure molecules of small or intermediate size. In this cases the use of hapten-radioimmunoassay is necessary, in which the material to be measured like steroid hormones, plant hormones, drugs and others, is coupled to a protein, usually albumin, and antibodies are raised in the usual manner to this protein-ligand conjugate. The assays of peptide hormones have been mostly done using iodinated tracers, but preferably in the hapten assays also other radionuclides are used. The iodination is carried out mostly with chloramine T, a strong oxidizing agent to couple radioactive iodine to tyrosyl or related residues of peptides [94–96]. For use the advantages of iodine labelled compounds (high specific activity, high sensitivity, convenient measurement) also for low molecular substances the hapten could be coupled to the protein via an iodinated bridge like tyramine. Thus a radioimmunoassay for digoxin is working with digoxin-(G-^{3}H), digoxin-(^{125}I)-tyrosine or digoxin-(^{125}I)-tyramine [97, 98] and an assay for the phytohormone abscisic acid is achieved with the tritium labelled compound [99]. In a semi-automated performance the assay permit the processing of several hundred samples per day.

After these analytical methods some other tracer methods will be mentioned. In each case by labelling endogenous substances with radioactive isotopes we can look very easy for the fate of these compounds. This can be topographically investigating the distribution of the applied compounds in plants and animals perhaps in dependence on the time. But it can be also for the research of the biotransformation of applied compounds or of the biosynthesis of endogenous substances. The use of radiolabelled compounds in drug metabolism studies has resulted in significant advances in this area of drug research preferably with phytoeffectors, herbicides, pestizides and others [64–69, 115, 122]. These studies are directed at an evaluation of all phases of drug metabolism, absorption, tissue distribution, metabolism, and excretion. In this cases sometimes the application of the labelled compounds or of precursors to the biological systems is difficult, preferably if the substance is only sparingly water soluble. At the end of the incubation period the compound under study has to be isolated from the biological system and purified to radiochemical homogenity. The identification of the metabolites can be carried out, for example, by reversed dilution analysis mentioned above. In biosynthetic experiments [100–103] following determination of the incorporation rate and/or dilution factor with endogenous pools, the final task often consists of establishing the location or distribution of the isotope in the isolated product by a chemical or biochemical degradation of the molecule into small fragments, often done in a stepwise fashion [1, 104], because it is not enough to have the result that the radioactivity of the applied precursor is found in the end product, this could happen also by a nonspecific pathway. Thus for example for the biosynthesis of nicotinic acid in plants after adding aspartic acid-4-^{14}C we are expecting nicotinic acid which contains all the radioactivity in the carboxylic group of this substance according to Fig. 4 [105].

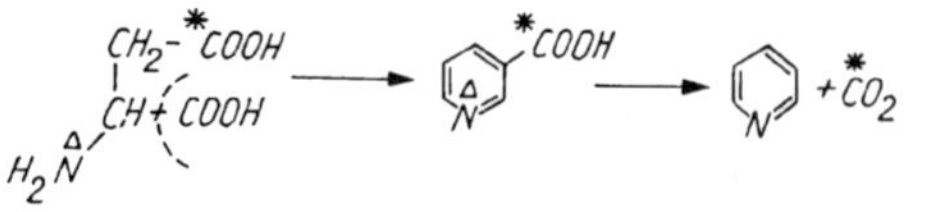

Fig. 4. Biosynthesis of nicotinic acid in plants

This has to be proved by degradation. If the CO_2 from the carboxylic group has the same specific radioactivity as the nicotinic acid then we can say that the whole radioactivity of the nicotinic acid is located in the carboxylic group. The experiments for degradation and for elucidation of the distribution of the radioactivity in the corresponding molecules are the most difficult work in biosynthetical research.

In recent years alternative approaches based on the use of stable isotopes, particularly ^{13}C, and employing spectroscopic methods, especially mass spectrometry and NMR spectroscopy, to locate the isotope in the product have become more and more prominent [102, 106, 107, 114].

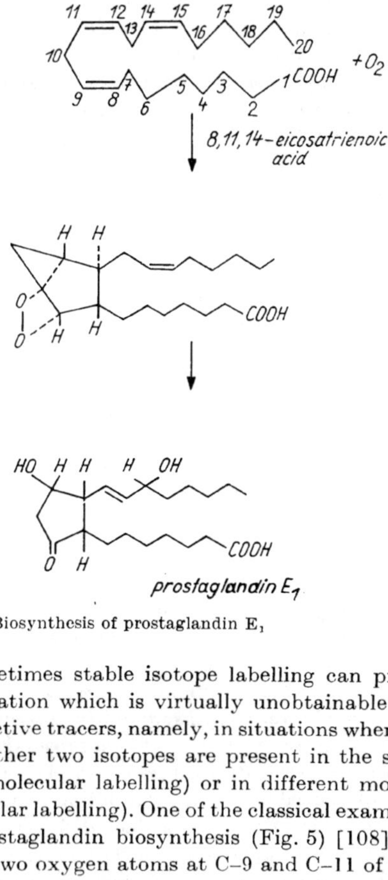

Fig. 5. Biosynthesis of prostaglandin E_1

Sometimes stable isotope labelling can provide certain information which is virtually unobtainable in work with radioactive tracers, namely, in situations where the question is whether two isotopes are present in the same molecule (intramolecular labelling) or in different molecules (intermolecular labelling). One of the classical examples is a study on prostaglandin biosynthesis (Fig. 5) [108] investigating if the two oxygen atoms at C-9 and C-11 of prostaglandin E_1 originate from the same molecule of oxygen or from two different O_2 molecules. For this, the biosynthesis of prostaglandin was carried out in an atmosphere of oxygen prepared by mixing about equal parts of normal oxygen (^{16}O-^{16}O) and > 99% ^{18}O enriched oxygen (^{18}O-^{18}O). If a molecule of oxygen is incorporated intact to give rise to the two oxygens at C-9 and C-11, the product will consist of species containing only ^{16}O atoms and species containing only ^{18}O atoms. If, on the other hand, the two oxygen atoms at C-9 and C-11 are derived from two different molecules of oxygen statistical recombination of ^{16}O and ^{18}O would occur and the product would contain 0,1 and 2 ^{18}O atoms molecule. The former is the case.

Recently, NMR spectroscopy and mass spectrometry is also used for the determination of the isotope distribution in radioactive labelled compounds [109–111]. Thus triton NMR has been involved in biosynthetic studies on penicillic acid [112]. More information in biosynthetic experiments is coming by use of multiple labelling techniques [103]. Thus for investigating the hypothesis that in the biosynthesis of nicotinic acid (Fig. 4) aspartic acid is incorporated without degradation double labelled (4-^{14}C, ^{15}N)-aspartic acid is used [113]. The same relation between ^{14}C-radioactivity and ^{15}N-enrichment in the precursor and in the final product shows the intact incorporation of the precursor into the nicotinic acid.

Another very important source of information about a biosynthetic pathway is the study of its stereochemistry, particularly by use of stereospecific labelled precursors [46]. Using the techniques mentioned a very broad knowledge has been obtained about the biosynthesis of different natural products like phenols, isoprenoids, alkaloids and others [116–121]. These are only few remarks about the application of radioactive isotopes in agricultural and related sciences.

Received February 13, 1980

References

[1] *H. R. Schütte*, Radioaktive Isotope in der organischen Chemie und Biochemie. VEB Deutscher Verlag der Wissenschaften, Berlin, 1966

[2] *E. A. Evans, M. Muramatsu* (Eds.), Radiotracer Techniques and Applications, Marcel Dekker, New York – Basel, 1977

[3] *G. v. Heresy, F. A. Paneth*, Z. anorg. Chem. **82** (1913) 323

[4] *E. Rutherford*, Nature (London) **103** (1919) 415

[5] *J. Curie, F. Joliot*, C. R. hebd. Séances Acad. Sci. **198** (1934) 254

[6] *E. Fermi, F. Amaldi, O. D'Agostino, F. Rasetti, E. Segré*, Proc. Roy. Soc. (London) **146** (1934) 483

[7] *O. Hahn, F. Strassmann*, Naturwissenschaften **27** (1939) 11

[8] *H. Koch*, Isotopenpraxis **15** (1979) 193

[9] *M. D. Kamen*, Science (Washington) **140** (1963) 584

[10] *L. Alvarez, R. Cornog*, Physic. Rev. **56**, (1939) 613

[11] International Atomic Energy Agency, Stable Isotopes in the Life Sciences. Proceedings of a technical committee meeting. Leipzig, 14.–18. 2. 1977, IAEA, Vienna 1977

[12] *E. Schönfeld*, Isotopenpraxis **14** (1978) 329

[13] *J. C. Clark, P. D. Buckingham*, Short-lived Radioactive Gases for Clinical Use, Butterworths, London 1975

[14] *H. Simon*, Messung von radioaktiven und stabilen Isotopen; Anwendung von radioaktiven und stabilen Isotopen in der Organischen Chemie und Biochemie. Band II, Springer-Verlag, Berlin-Heidelberg 1974

[15] *E. D. Bransome*, Jr. (Ed.), The Current Status of Liquid Scintillation Counting. Grune and Stratton, New York 1970

[16] *P. Stanley, B. Scoggins* (Eds.), Liquid Scintillation Counting-Recent Advances, Academic Press, New York 1974

[17] *Y. Kobayashi, D. V. Maudsley*, Biological Applications of Liquid Scintillation Counting, Academic Press, New York 1974

[18] *D. L. Horrocks*, Applications of Liquid Scintillation Counting. Academic Press, New York 1974

[19] *M. A. Crook, P. Johnson* (Eds.), Liquid Scintillation Counting. Vol. 4, Heyden and Son Ltd., London 1976

[20] *B. W. Fox*, Techniques of Sample Preparation for Liquid Scintillation Counting. North Holland, Amsterdam 1976

[21] *C. T. Peng*, Sample preparation in liquid scintillation counting. Review No. 17, The Radiochemical Centre, Amersham, England, 1977

[22] *K. Buchtela, M. Tschurlovits*, Int. J. Appl. Radiat. Isotop. **26** (1975) 333

[23] *F. Fric, V. Finocchiaro*, Radiochem. Radioanal. Lett. **21** (1975) 205

[24] *W. J. Gelsema, C. L. Ligny, J. B. Luter, F. G. A. Vossenberg*, Int. J. Appl. Radiat. Isotop. **26** (1975) 443

[25] *J. G. T. Regan, J. F. C. Tyler*, Analyst (London) **101** (1976) 32

[26] *R. M. Fink, C. E. Dent, K. Fink*, Nature (London) **160** (1947) 801

[27] *S. Prydz*, Anal. Chem. **45** (1973) 2317

[28] *U. Lüthi, P. G. Waser*, Nature (London) **205** (1965) 1190

[29] *J. F. Koren, T. B. Melo, S. Prydz*, J. Chromatogr. **46**, (1970) 129

[30] *P. B. Gahan* (Ed.), Autoradiography for Biologists, Academic Press, London 1972

[31] *A. W. Rogers*, Techniques of Autoradiography, 2nd ed., Elsevier, Amsterdam 1973

[32] *E. Amlacher*, Autoradiographie in Histologie und Zytologie. VEB Georg Thieme, Leipzig 1974

[33] *L. E. Feinendegen*, Prospects and Problems in the Application of Isotopes to Pure and Applied Cytology, in: [2] Vol. 2 (1977) 769

[34] *L. Benes*, Isotopenpraxis **14** (1978) 335

[35] *A. Murray, D. L. Williams*, Organic Syntheses with Isotopes. Part I. Compounds of Isotopic Carbon. Wiley-Interscience, New York 1958

[36] *A. Murray, D. L. Williams*, Organic Syntheses with Isotopes. Part II. Organic Compounds Labelled with Isotopes of the Halogens, Hydrogen, Nitrogen, Oxygen, Phosphorus, and Sulfur. Wiley-Interscience, New York 1958

[37] *M. Wenzel, P. E. Schulze*, Tritium-Markierung. De Gruyter, Berlin 1962

[38] *M. Bubner, L. Schmidt*, Die Synthese Kohlenstoff-14-markierter organischer Verbindungen. Georg Thieme, Leipzig 1966

[39] *J. R. Catch*, Preparation of Radioactive Tracers: Synthetic Organic Tracers. International Encyclopaedia of Pharmacology and Therapeutics, Section 78, Vol. 1, pp. 97–130; Radionuclides in Pharmacology, Pergamon Press, Oxford 1972

[40] *E. A. Evans*, Tritium and its Compounds. 2nd ed. Butterworths, London 1974

[41] *H. R. Schütte, G. Bachmann*, Chem. Techn. **12** (1960) 599

[42] *R. F. Dawson, D. R. Christman, A. D'Adamo, M. L. Solt, A. P. Wolf*, J. Amer. chem. Soc. **82** (1960) 2628

[43] *E. A. Evans, H. C. Sheppard, J. C. Turner, D. C. Warrell*, J. Label. Compounds **10** (1974) 569

[44] *G. W. Kirby, L. Ogunkoya*, J. chem. Soc. (London) **1965**, 6914

[45] *N. Johns, G. W. Kirby*, Chem. Commun. **1971, 163**

[46] *H. R. Schütte, M. Stock*, Isotopenpraxis **15** (1979) 197

[47] *K. C. Tovey, G. H. Spiller, K. G. Oldham, N. Lucas, N. G. Carr*, Biochem. J. **142** (1974) 47

[48] *K. E. Wilzbach*, J. Amer. Chem. Soc. **79** (1957) 1013

[49] *A. Wacker, L. Träger*, Angew. Chem. **72** (1960) 168

[50] *E. A. Evans*, Self-decomposition of radiochemicals, Review No. 16, The Radiochemical Centre, Amersham, England, 1976

[51] *E. A. Evans*, Storage and Stability of Radiotracer Compounds, in: [2] Vol. 1 (1977) 237

[52] *M. Wolfsberg*, Ann. Rev. Phys. Chem. **20** (1969) 449

[53] *C. J. Collins, N. S. Bowman* (Eds.), Isotope Effects in Chemical Reactions, Van Nostrand, New York 1971

[54] *R. L. Schowen*, Progr. Phys. Org. Chem. **9** (1972) 275

[55] Radiation and Radioisotopes Applied to Insects of Agricultural Importance, IAEA, Vienna 1963

[56] Radiation and Radioisotopes for Industrial Microorganisms, IAEA, Vienna 1971

[57] *D. J. Nelson* (Ed.), Radionuclides in Ecosystems, *U. S.* Atomic Energy Commission, Oak Ridge, Tenn. 1971

[58] Nuclear Techniques in Environmental Pollution, IAEA, Vienna 1971

[59] *E. Gaspar, M. Oncescu*, Radioactive Tracers in Hydrology, Elsevier, New York 1972

[60] *A. A. Moghissi, M. W. Carter*, Radionuclides in Environmental Studies, in: [2], Vol. 1 (1977) 659

[61] Radioisotopes in Hydrology, IAEA, Vienna 1963

[62] *E. C. Tsivoglou, J. B. Cohen, S. D. Shearer, P. J. Godsil*, J. Water Poll. Cont. Fed. **40** (1968) 285

[63] *E. Soini*, Science Tools **25** (1978) 38 (The LKB Instrument Journal)

[64] *F. M. Ashton, A. S. Crafts*, Mode of Action of Herbicides, Dekker, New York 1973

[65] *L. J. Audus*, Physiology and Biochemistry of Herbicides. Academic Press, London 1974

[66] *P. C. Kearney, D. D. Kaufman* (Eds.), Herbicides, Chemistry, Degradation and Mode of Action. Dekker, New York, Vol. I and II, 1976

[67] *G. G. Still, G. Paulson, D. R. Nelson*, The Use of Radioisotopes to the Study of the Metabolism of Agricultural Chemicals, in: [2], Vol. 2 (1977) 893

[68] *H. R. Schütte*, in: *T. Kudrev, I. Ivanova, E. Karanov*, Plant Growth Regulators, Bulgarian Academy of Sciences, Sofia 1977, p. 387

[69] *H. Gräser*, Biochemie und Physiologie der Phytoeffektoren, VEB Deutscher Verlag der Wissenschaften, Berlin 1978

[70] *L. F. Chasseaud, D. R. Hawkins, B. D. Cameron, B. J. Fry, V. H. Saggers*, Xenobiotica **2** (1972) 269

[71] *Y. Nakai, Y. Shirakawa, T. Fujita, Z. Suzuoki*, Xenobitoca **2** (1972) 147

[72] *E. H. Belcher, H. Vetter* (Eds.), Radioisotopes in Medical Diagnosis. Butterworths, London 1971
[73] *J. A. DeGrazia*, Diagnostic Applications of Radionuclides, in: [2], Vol. 2 (1977) 1005
[74] *V. R. McCready, D. M. Taylor, N. G. Trott, C. B. Cameron, E. O. Field, R. J. French, R. P. Parker* (Eds.), Radioactive Isotopes in the Localization of Tumours. Heinemann, London 1969
[75] *J. S. Mitchell*, Radioactive Drugs in Cancer Therapy, in: [2] Vol. 2 (1977) 1081
[76] *K. G. Oldham*, Radiochemical Methods in Enzyme Assay, Review No. 9, The Radiochemical Centre, Amersham, England, 1968
[77] *K. G. Oldham*, J. Label. Compounds 4 (1968) 127
[78] *K. G. Oldham*, Anal. Biochem. **44** (1971) 143
[79] *K. G. Oldham*, Methods Biochem. Anal. **21** (1973) 191
[80] *K. G. Oldham*, Radiotracer Techniques for Enzyme Assays and Enzymatic Assays, in: [2] Vol. 2 (1977) 823
[81] *M. Sugii, H. Miura, K. Nagata*, Anal. Biochem. **92** (1979) 265
[82] *J. K. Saelens, M. S. Schoen, G. B. Kovacsis*, Biochem. Pharmacol. **16** (1967) 1043
[83] *A. C. Cuello, A. S. Horn, A. V. P. Mackay, L. L. Iversen*, Nature (London) **243** (1973) 465
[84] *J. B. Lombardini, M. K. Burch, P. Talalay*, J. Biol. Chem. **246** (1971) 4465
[85] *I. B. Rubin, G. Goldstein*, Anal. Biochem. **33** (1970) 244
[86] *S. Spector*, Ann. Rev. Pharmacol. **13** (1973) 959
[87] *P. H. Sönksen*, Br. Med. Bull. **30** (1974) 1
[88] *C. A. Pasternak* (Ed.), Radioimmunoassay in Clinical Biochemistry, Heyden and Son Ltd., London 1975
[89] *H. G. Eckert*, Angew. Chem. 88 (1976) 565
[90] *J. Landon, A. C. Moffat*, Analyst **101** (1976) 225
[91] *B. E. P. Murphy, D. C. Lehotay*, Competitive Protein-Binding Assays, in: [2] Vol. 2 (1977) 969
[92] Radioimmunoassay and Related Procedures in Medicine. Proceedings of Symposium, Berlin 31. 10. – 4. 11. 1977. IAEA, Vienna 1977
[93] *R. S. Yalow, S. A. Berson*, Nature (London) **184** (1959) 1648
[94] *F. C. Greenwood, W. M. Hunter, J. S. Glover*, Biochem. J. **89** (1963) 114
[95] *H. E. Schmidt, B. Teichmann, R. Vogt, H. Herzmann*, Isotopenpraxis **10** (1974) 401
[96] *A. E. Bolton*, Radioiodination Techniques. Review No. 18, The Radiochemical Centre, Amersham, England, 1977
[97] *E. W. Weiler, M. H. Zenk*, Phytochemistry **15** (1976) 1537
[98] *E. W. Weiler, M. H. Zenk*, Clin. Chem. **25** (1979) 44
[99] *E. W. Weiler*, Planta **144** (1979) 255
[100] *S. A. Brown*, in Biosynthesis, Vol. 1, Specialist Periodical Reports (*T. A. Geissman*, ed.), The Chemical Society, London 1972
[101] *S. A. Brown, L. R. Wetter*, Progress in Phytochemistry **3** (1972) 1
[102] *H. G. Floss*, Lloydia **35** (1972) 399
[103] *H. G. Floss*, Radiotracers in Biosynthetic Studies, in [2] Vol. 2 (1977) 689
[104] *H. Simon, H. G. Floss*, Bestimmung der Isotopenverteilung in markierten Verbindungen. Springer-Verlag, Berlin 1967
[105] *E. Mothes, D. Gross, H. R. Schütte, K. Mothes*, Naturwissenschaften **48** (1961) 623
[106] *J. B. Grutzner*, Lloydia **35** (1972) 375
[107] *U. Sequin, A. I. Scott*, Science **186** (1974) 101
[108] *B. Samuelsson*, J. Amer. Chem. Soc. **87** (1965) 3011
[109] *J. M. A. Al-Rawi, J. P. Bloxsidge, J. A. Elvidge, J. R. Jones, V. M. A. Chambers, E. A. Evans*, J. Label. Comp. and Radiopharmaceuticals **12** (1976) 293
[110] *J. P. Bloxsigde, J. A. Elvidge, J. R. Jones, R. B. Mane, V. M. A. Chambers, E. A. Evans, D. Greenslade*, J. Chem. Res. **1977**, 42
[111] *V. M. A. Chambers, E. A. Evans, J. A. Elvidge, J. R. Jones*, Tritium Nuclear Magnetic Resonance (tnmr) Spectroscopy, Review, No. 19, The Radiochemical Centre, Amersham, England, 1978
[112] *J. M. A. Al-Rawi, J. A. Elvidge, D. K. Jaiswal, J. R. Jones, R. Thomas*, Chem. Commun. **1974**, 220
[113] *D. Gross, H. R. Schütte, G. Hübner, K. Mothes*, Tetrahedron Letters (London) **1968**, 541
[114] *E. Buncel, C. C. Lee*, Carbon-13 in Organic Chemistry; Isotopes in Organic Chemistry. Vol. 3, Elsevier, Amsterdam 1977
[115] *H. R. Schütte* (Ed.), Wirkungsmechanismus von Herbiziden und synthetischen Wachstumsregulatoren. VEB Gustav Fischer Verlag, Jena 1980
[116] *P. Bernfeld* (Ed.), Biogenics of Natural Compounds, Pergamon Press, Oxford 1967
[117] *J. D. Bu'Lock*, Biosynthese von Naturstoffen. Eine Einführung in den Sekundärstoffwechsel, Bayrischer Landwirtschaftsverlag, München 1970
[118] *M. Luckner*, Sekundärstoffwechsel in Pflanze und Tier. VEB Gustav Fischer Verlag, Jena 1969
[119] *K. Mothes, H. R. Schütte* (Eds.), Biosynthese der Alkaloide. VEB Deutscher Verlag der Wissenschaften, Berlin 1969
[120] *H. R. Schütte*, Progr. in Botany **33** (1971) 105; **34** (1972) 165; **35** (1973) 103; **36** (1974) 108; **37** (1975) 133; **38** (1976) 129; **39** (1977) 83; **40** (1978) 126; **41** (1979) 93
[121] *H. R. Schütte*, Biosynthese von Naturstoffen. VEB Gustav Fischer Verlag, Jena 1981
[122] *H. R. Schütte*, Isotopenpraxis (in Vorbereitung)
[123] *R. M. Schultz, P. M. Wassarman*, Anal. Biochem. **77** (1977) 25
[124] *J. Airhart, J. Kelley, J. E. Brayden, R. B. Low*, Anal. Biochem. **96** (979) 45
[125] *H. G. Eckert, H. Strecker*, Naturwissenschaften **65** (1978) 245
[126] *M. F. L'Annunziata*, Radiotracers in Agricultural Chemistry. Academic Press, New York 1979

Fortschrittsberichte

STAND UND ENTWICKLUNGSTENDENZEN AUF DEM GEBIET DER STRAHLENSICHERHEIT UND DES UMWELTSCHUTZES

DOSIMETRISCHE VERFAHREN UND GERÄTE

Lyolumineszenzdosimetrie (LLD)

W. Stolz

(Bergakademie Freiberg, Sektion Physik)[1])

Inhaltsübersicht

Keywords

chemiluminescense; dosemeters; dosimetry; luminescent dosemeters; organic compounds; radiation doses; reviews

1. Einführung

Das Phänomen der Lyolumineszenz wurde erstmals 1895 von *Wiedemann* und *Schmidt* an Alkalihalogeniden nach Bestrahlung mit Katodenstrahlen beobachtet. In ihrer Arbeit „Über Luminescenz", die in den „Annalen der Physik und Chemie" erschien, schreiben sie [24]:

> „Ferner haben wir eine weitere Art der Luminescenz aufgefunden, welche man mit dem Namen Lyoluminescenz belegen könnte. In gewissen Fällen tritt beim Lösen von Substanzen eine Lichtentwicklung auf, nachdem sie vorher bestrahlt worden sind, während sie eine solche ohne Bestrahlung nicht zeigen".

Erst 60 Jahre später wurde das Studium dieser Lumineszenzerscheinung erneut aufgegriffen. *Ahnström* und *Ehrenstein* fanden 1959, daß auch beim Auflösen bestrahlter organischer Verbindungen (Glukose) Licht ausgestrahlt wird [1]. Ausgangspunkt der Deutung der Lyolumineszenz anorganischer Substanzen bilden Untersuchungen von *Ahnström* an bestrahltem NaCl [2]. Auf die Möglichkeit der Anwendung der Lyolumineszenz zur integrierenden Dosimetrie ionisierender Strahlung haben zuerst *Atari, Ettinger* und *Fremlin* 1973 hingewiesen [4].

2. Mechanismus der Lyolumineszenz

Alkalihalogenide

Der Lyolumineszenzprozeß der Alkalihalogenide wurde eingehend am NaCl nach Bestrahlung mit γ-Strahlung untersucht. Als Lösungsmittel dienten reines Wasser, Fluoresceinlösung und TlCl-Lösung [2, 4, 7, 8, 16, 23]. Die Diskussionen über den Mechanismus der Lyolumineszenz sind noch nicht abgeschlossen. Übereinstimmung herrscht aber darüber, daß die Lichtemission maßgeblich auf die Rekombination der durch die Bestrahlung gebildeten F- und V_2-Zentren zurückzuführen ist. Es kann als sehr wahrscheinlich angenommen werden, daß in reinem Wasser ein anderer Lumineszenzmechanismus vorliegt als in der wäßrigen Lösung eines fluoreszierenden Stoffes.

Nach *Ettinger, Atari* u. Mitarb. findet in reinem Wasser die Emission unmittelbar an der Oberfläche der sich auflösenden Kristalle statt [16]. Beim Lösungsvorgang entstehen hydratisierte Elektronen e^-_{aq} (aus F-Zentren) und Halogenatome (aus V_2-Zentren). Die Lyolumineszenz wird dann durch folgende Reaktionsschritte gekennzeichnet:

$$e^-_{aq} + e^-_{aq} \rightarrow H_2 + 2\,OH^-$$

$$e^-_{aq} + \dot{Cl} \rightarrow \overset{*}{Cl^-} \rightarrow Cl^- + hf$$

Teteris beschreibt ausführlich die Lyolumineszenz beim Auflösen bestrahlter Alkalihalogenide in TlCl-Lösung [23]. Er gelangt zu der Ansicht, daß die Emission auf die Rekombination der F- und V_2-Zentren in der Oberflächenschicht der Kristalle zurückzuführen ist. Es werden Halogenionen in angeregten Triplett-Zuständen gebildet, die in der Lösung die Anregungsenergie an fluoreszenzfähige Moleküle oder Löschmoleküle abgeben.

Die von *Avotinsh, Gorbovicka* und *Tiliks* beobachtete Lyolumineszenz von bestrahltem LiF in wäßriger Flouresceinlösung steht ebenfalls in enger Beziehung zur Bildung von F-Zentren [9].

Organische Verbindungen

Unter den organischen Stoffen wurde die Lyolumineszenz von Sacchariden nach Bestrahlung am häufigsten untersucht. *Atari, Ettinger* und Mitarb. haben gezeigt, daß in wäßriger Lösung Glukose, Glukosemonohydrat, Mannose,

[1]) Anschrift: DDR-9200 Freiberg, Bernhard-von-Cotta-Straße 4

Xylose und Trehalosedihydrat das stärkste Lyolumineszenzvermögen aufweisen [5, 6, 12, 13]. Es wird angenommen, daß für die Lumineszenzemission bei der Auflösung der bestrahlten Stoffe Reaktionen freier Radikale im wäßrigen Milieu in Betracht kommen. Folgender Lyolumineszenz-Reaktionsmechanismus wird vorgeschlagen:

$$R\cdot + R\cdot \rightarrow R_2$$

$$R + O_2 \rightarrow RO_2^{\cdot}$$

$$RO_2^{\cdot} + RO_2^{\cdot} \rightarrow R_2 + (O_2)_2^*$$

$$(O_2)_2^* \rightarrow (O_2)_2 + hf$$

Eine um Größenordnungen verstärkte Lumineszenzemission wird beobachtet, wenn man anstatt Wasser Lösungen der bekannten chemilumineszenzfähigen Verbindungen Luminol (5-Amino-1, 2, 3, 4-tetrahydrophthalazindion-1, 4) oder Lucigenin (N, N'-Bis-methylacridinium-dinitrat) als Lösungsmittel für die bestrahlten Substanzen verwendet [12, 15, 22].

In Luminollösungen wurden folgende Verstärkungsfaktoren der Lichtemission ermittelt [12]:

Glukose:	$6{,}7 \cdot 10^5$
Xylose:	$5{,}5 \cdot 10^4$
Mannose:	$5{,}8 \cdot 10^3$
Trehalose · 2 H_2O:	$3{,}0 \cdot 10^3$

Ebenso wie bei einigen Sacchariden werden auch bei einer Reihe organischer Verbindungen anderer Stoffklassen Lyolumineszenzreaktionen mit größerer Lichtausbeute beobachtet. *Ettinger*, *Rowe* und Mitarb. berichten über systematische Untersuchungen an zahlreichen Aminosäuren, Proteinen und Nukleinsäuren [14, 17]. Durch ein stärkeres Lyolumineszenzvermögen zeichnen sich Glutamin, Glutaminsäure, Theronin, Valin und Phenylalanin aus.

Für die Dosimetrie ionisierender Strahlung sind die Ergebnisse der Untersuchungen von *Alechin*, *Babenko*, *Keirim-Markus* und Mitarb. bemerkenswert [3]. Sie fanden, daß auch bestrahlte körpereigene Substanzen (Hornhaut, Nägel, Haare) beim Auflösen in Natriumsulfidlösung lyolumineszieren.

3. Lyolumineszenzmessung

Bei der Untersuchung der Lyolumineszenz verfolgt man den Vorgang der Lichtemission während der Auflösung bestrahlter Festkörper in einem Lösungsmittel. Der Lumineszenzlichtstrom Φ_{LL} wird als Funktion der Zeit t aufgezeichnet. Die Abb. 1 zeigt typische Beispiele für den Verlauf der Funktion $\Phi_{LL} = \Phi_{LL}(t)$. Bei der dosimetrischen Anwendung der Lyolumineszenz dient das Zeitintegral des Lichtstromes, die Lichtsumme

$$Q_{LL} = \int_{t_1}^{t_2} \Phi_{LL}(t)\, dt$$

als Maß für die Energiedosis. Dabei sind t_1 und t_2 Anfangs- und Endzeitpunkt des Auflösungsvorganges. Eine einfache Anordnung zur Messung der Lyolumineszenz besteht daher aus folgenden Teilen: Küvette, Photovervielfacher, Schreiber bzw. Integrator. Eine schnelle Auflösung

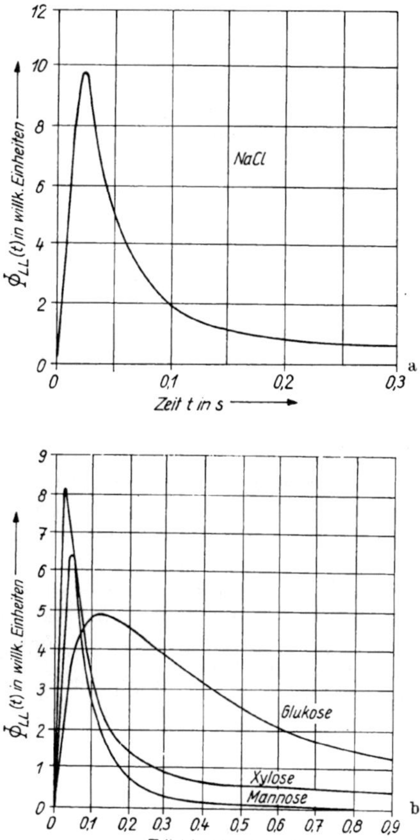

Abb. 1. Zeitlicher Verlauf des Lichtstromes der Lyolumineszenz bestrahlter Substanzen (Lösungsmittel: Wasser)
(a) NaCl (nach [4])
(b) Glukose, Xylose, Mannose (nach [5])

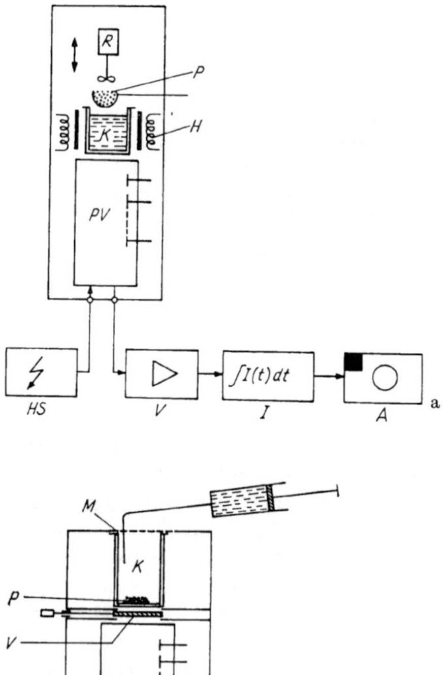

Abb. 2. Schematische Darstellung von Anordnungen zur Messung der Lyolumineszenz (nach [7, 21])
PV – Photovervielfacher; K – Küvette; HS – Hochspannung; P – Probe; V – Verstärker; V – Verschluß; I – Integrator; H – Heizer; A – Anzeige; M – Gummimembran

der Substanz (10 bis 50 mg) wird erreicht, wenn das Lösungsmittel (ca. 5 ml) gerührt und mittels eines Heizers auf erhöhter Temperatur gehalten wird. Zwei übliche Meßanordnungen sind in Abb. 2 schematisch dargestellt. Die Substanz wird entweder mittels einer Dosiervorrichtung in das Lösungsmittel eingebracht (Abb. 2 a), oder das Lösungsmittel auf die in der Küvette befindliche Probe geleitet (Abb. 2 b).

Im Hinblick auf die oftmals sehr geringe Lichtausbeute schlagen *Bartlett* und *Kannan* das Verfahren der Photonenzählung für Lyolumineszenzmessungen vor [10, 18].

4. Dosimetersubstanzen und ihre Eigenschaften

Dosimetersubstanzen

Von den vielen Substanzen, die nach Bestrahlung Lyolumineszenz zeigen, sind nur wenige für die Dosimetrie geeignet. Eingehende Untersuchungen der dosimetrischen Eigenschaften sind bisher erst an NaCl [4, 7, 8], Mannose [5, 10, 12, 13, 18, 20, 21, 22], Glukose [5, 11–13, 19, 22] und Trehalosedihydrat [6, 12, 13, 22] durchgeführt worden. Dabei richtet sich wegen des annähernd gewebeäquivalenten Verhaltens das Hauptaugenmerk auf die organischen Verbindungen. Es hat sich gezeigt, daß bereits Spuren von Verunreinigungen die Lyolumineszenzemission beträchtlich verändern. Besondere Bedeutung kommt dem Vorschlag zu, die Lyolumineszenz körpereigener Substanzen zur Abschätzung der Strahlenbelastung nach Havarien heranzuziehen [3].

Dosismeßbereich

Die Dosischarakteristik einer lyolumineszierenden Substanz stellt die Abhängigkeit der Lichtsumme Q_{LL} von der Energiedosis D dar. Sie wird durch die Lumineszenzausbeute

$$\eta_{LL} = \frac{W_{LL}}{mD}$$

bestimmt, wobei W_{LL} die von einer Probe der Masse m emittierte Lichtenergie und D die in der Substanz durch Bestrahlung erzeugte Energiedosis D bedeuten. Wenn η_{LL} konstant ist, verläuft die Dosischarakteristik linear:

$$Q_{LL} = k\eta_{LL}\, D$$

(k – Proportionalitätskonstante). Die untere Grenze D_u des Meßbereiches hängt von den Dunkelstromschwankungen des Photovervielfachers ab. Bei Verwendung von Luminollösung macht sich außerdem die Untergrundlumineszenz infolge von Chemilumineszenzreaktionen mit gelöstem Sauerstoff und oxydierenden Verunreinigungen störend bemerkbar. Typische Dosischarakteristiken zeigen die Abb. 3 a bis 3 c.

Bei NaCl erstreckt sich der gesamte ausnutzbare Energiedosismeßbereich von etwa 10 mGy bis 100 kGy (Meßunsicherheit: ± 5%). Proben verschiedener Herkunft zeigen starke Empfindlichkeitsunterschiede [4, 7].

Unter den organischen lyolumineszierenden Substanzen ist Mannose am besten hinsichtlich der Eignung als Dosimetersubstanz untersucht worden. Die Dosischarakeristik von Mannose verläuft zwischen 50 mGy und 3 Gy linear, darüber bis 1,2 kGy supralinear.

Ein Einfluß der Energiedosisleistung auf die Lumineszenzausbeute war im Bereich von etwa 0,05 Gy/s bis 100 Gy/s nicht feststellbar. Die Meßunsicherheit liegt zwischen ± 3 und ± 4% [20, 21].

Mit Trehalosedihydrat (Abb. 3 c) haben *Atari* und *Ettinger* Energiedosen zwischen 1 mGy und 1,5 kGy erfaßt [6, 12]. *Puite* und *Crebolder* konnten jedoch die hohe Empfindlichkeit dieses Saccharids nicht bestätigen [21].

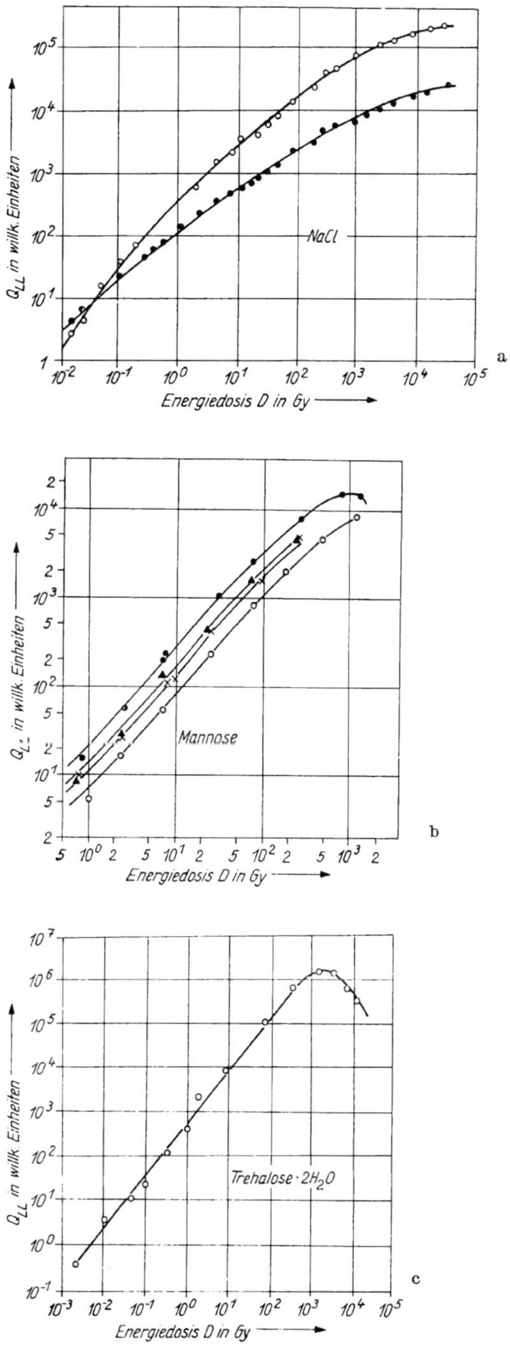

Abb. 3. Typische Energiedosis-Anzeigecharakteristiken einiger lyolumineszierender Materialien (Lösungsmittel: Wasser)
(a) NaCl, verschiedener Herkunft; ^{60}Co-γ-Strahlung (nach [7])
(b) Mannose (nach [21]); ● – ^{60}Co-γ-Strahlung; △ – 15 MeV-Neutronen; × – 5,3 MeV-Neutronen; ○ – Spaltungsneutronen
(c) Trehalosedihydrat, ^{60}Co-γ-Strahlung (nach [6])

γ-Energieabhängigkeit

Organische lyolumineszierende Substanzen verhalten sich gegenüber Photonenstrahlung annähernd gewebeäquivalent. Bei kleinen Photonenenergien ($W_\gamma < 100$ keV) zeigen Saccaride im Vergleich zu Muskelgewebe erwartungsgemäß einen unterschwelligen Verlauf der Energieab-

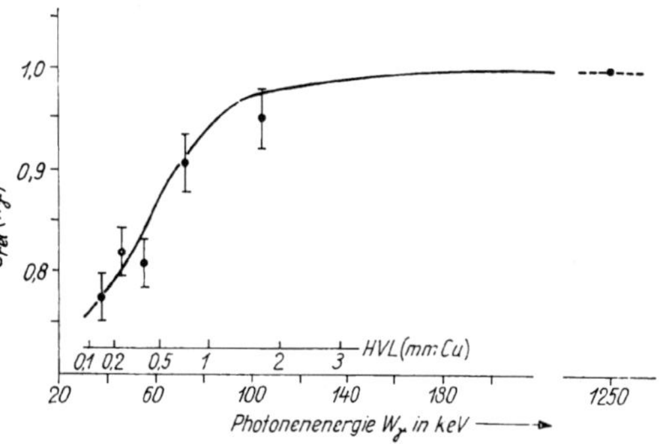

Abb. 4. Berechnete und gemessene γ-Energieabhängigkeit von Mannose (nach [21])

hängigkeitskurve. Abb. 4 gibt die Energieabhängigkeit der relativen Meßempfindlichkeit

$$\varepsilon_{\text{rel}}(W_\gamma) = \frac{\left[\frac{\mu_{\text{en}}(W_\gamma)}{\varrho}\right]_{\text{Mannose}}}{\left[\frac{\mu_{\text{en}}(W_\gamma)}{\varrho}\right]_{\text{Muskel}}} : \frac{\left[\frac{\mu_{\text{en}}(1{,}25\ \text{MeV})}{\varrho}\right]_{\text{Mannose}}}{\left[\frac{\mu_{\text{en}}(1{,}25\ \text{MeV})}{\varrho}\right]_{\text{Muskel}}}$$

von Mannose in bezug auf Standard-Muskelgewebe wieder.

Neutronenempfindlichkeit

Die meisten Festkörperdosimeter (z. B. Thermolumineszenzdosimeter) besitzen für schnelle Neutronen nur eine sehr geringe Empfindlichkeit. Saccharide eröffnen daher wegen ihrer hohen Lyolumineszenzausbeute und ihrer annähernd gewebeähnlichen Zusammensetzung gute Perspektiven für die Neutronendosimetrie. Allerdings unterscheiden auch sie sich im Wasserstoffgehalt (≈ 7 Gew.%) noch merklich von Muskelgewebe (10,2 Gew.%). Bei Bestrahlung in verschiedenen Neutronenfeldern wurden von *Puite* für Mannose die in Tab. 1 zusammengestellten Empfindlichkeiten, bezogen auf die Empfindlichkeit gegenüber ^{60}Co-γ-Strahlung, ermittelt [20]. Die Empfindlichkeit ist dabei definiert als Quotient aus Meßeffekt und Gewebeenergiedosis. Es wird vorgeschlagen, durch Mischen der lyolumineszierenden organischen Substanzen mit Polyäthylenpulver (Erhöhung des Wasserstoffgehaltes) die Empfindlichkeit für schnelle Neutronen anzuheben [20, 21].

Tab. 1. Relative Empfindlichkeit $\varepsilon_n/\varepsilon_\gamma$ von Mannose für schnelle Neutronen (nach [20])

Strahlung	$\varepsilon_n/\varepsilon_\gamma$
Spaltungsneutronen	0,24 ± 0,02
5,3 MeV-Neutronen	0,45 ± 0,05
15 MeV-Neutronen	0,60 ± 0,04

Fading

Für die praktische Dosimetrie ist die Abnahme des Lyolumineszenzvermögens während der Lagerungszeit zwischen Bestrahlung und Auswertung (Fading) von Bedeutung. Fading und Lumineszenz (Afterglow) der bestrahlten Substanzen im festen Zustand stehen in enger Beziehung.

Das Fading von NaCl beschreiben *Atari* und *Ettinger* [7, 8]. Es hängt von der Energiedosis ab. Nach 7wöchiger Lagerung bei 20 °C wurde eine Verringerung des Lyolumineszenzvermögens um 15% ($D = 4{,}6$ kGy) bzw. 30% ($D = 53$ Gy) beobachtet. Zur Erklärung dieser Erscheinung wird angenommen, daß bei hohen Energiedosen bevorzugt V_3-Zentren gebildet werden, die eine größere Stabilität als V_2-Zentren besitzen.

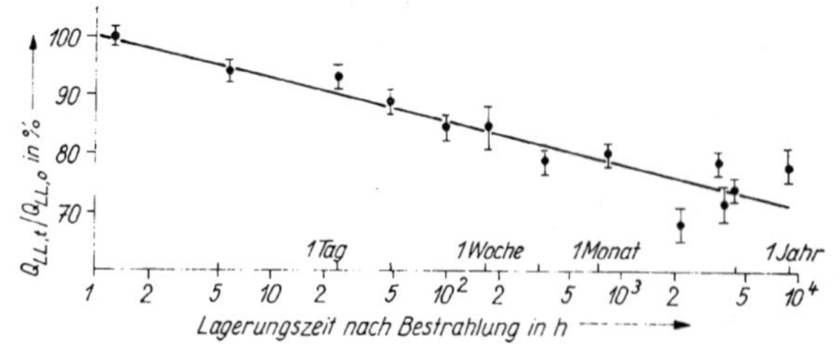

Abb. 5. Mannose: Fading der Energiedosisanzeige in Abhängigkeit von der Lagerungszeit bei Zimmertemperatur (nach [21])

Fadingwerte für Mannose werden von einigen Autoren angegeben [10, 13, 21]. Eine Fadingkurve von Mannose zeigt die Abb. 5. Über eingehende Untersuchungen des Fadingverhaltens dieses Saccharids berichtet *Bartlett* [10]. Er stellte fest, daß sich das Lyolumineszenzvermögen innerhalb von 6 Stunden nach Bestrahlungsende um ca. 10% verringert und in den folgenden 240 Tagen von 90% auf 60% abklingt, wenn die Energiedosis $D < 0{,}5$ kGy bleibt. Bei Lagerung nach Bestrahlung mit hohen Energiedosen ($D > 1$ kGy) beobachtete er hingegen ein Anwachsen der Lyolumineszenzfähigkeit. Für dieses Verhalten wird die örtliche Sauerstoffkonzentration in der Nähe der gebildeten freien Radikale verantwortlich gemacht.

Anwendung von Lyolumineszenzdosimetern

Gegenwärtig ist die Lyolumineszenzdosimetrie noch nicht so weit ausgereift, daß sie nennenswerten Eingang in die praktische Dosimetrie gefunden hat. Die Dosimetersubstanzen sind jedoch billiger als andere Festkörperdosimeter. Auswertegeräte für Lyolumineszenzdosimeter lassen sich darüber hinaus mit geringem technischen und elektronischen Aufwand realisieren. Für die Messung schneller Neutronen und niederenergetischer Photonenstrahlung dürfte die gewebeähnliche Zusammensetzung einiger lyolumineszierender organischer Verbindungen sehr vorteilhaft sein. Lyolumineszenzdosimeter könnten daher für die Havariedosimetrie, die klinische Dosimetrie, die strahlenbiologische Dosimetrie und die strahlentechnische Dosimetrie Bedeutung gewinnen.

Eingegangen am 17.3.1980

Literatur

[1] *G. Ahnström, G. Ehrenstein*, Acta Chem. Scand. **13** (1959) 855

[2] *G. Ahnström*, Acta Chem. Scand. **19** (1965) 300

[3] *I. A. Alechin, S. P. Babenko, I. B. Keirim-Markus, S. N. Kraitor, K. K. Kušnereva*, Atomn. Energ. **44** (1978) 347

[4] *N. A. Atari, K. V. Ettinger, J. H. Fremlin*, Radiat. Effects **17** (1973) 45

[5] *N. A. Atari, K. V. Ettinger*, Radiat. Effects **20** (1973) 135

[6] *N. A. Atari, K. V. Ettinger*, Nature **249** (1974) 341

[7] *N. A. Atari, K. V. Ettinger*, Proc. of the Fourth Int. Conf. on Luminescence Dosimetry, Krakow 1974, Vol. 2, 395 (Ed. *T. Niewiadomski*)

[8] *N. A. Atari, K. V. Ettinger*, Radiat. Effects **26** (1975) 39

[9] *Ju. E. Avotinsh, T. I. Gorbovicka, Ju. E. Tiliks*, Izv. Akad. Nauk Latv. SSR, Ser. Fiz. **1978**, 122

[10] *D. T. Bartlett*, Phys. Med. Biol. **24** (1979) 63

[11] *J. S. Chazhoor, U. C. Mishra*, J. Luminescence **20** (1979) 397

[12] *K. V. Ettinger, N. A. Atari*, Proc. of the Fourth Int. Conf. on Luminescence Dosimetry, Krakow 1974, Vol. 2, 417 (Ed. *T. Niewiadomski*)

[13] *K. V. Ettinger, J. R. Mallard, S. Srirath, A. Takavar*, Phys. Med. Biol. **22** (1977) 481

[14] *K. V. Ettinger, R. W. Rowe, J. R. Mallard, A. Takavar, J. Sephton*, Proc. of the Fourth Int. Congr. Radiat. Prot. Ass., Paris 1977, 1265

[15] *K. V. Ettinger, A. Takavar, J. R. Mallard, G. Buchan*, Proc. of the Fourth Int. Congr. Radiat. Prot. Ass., Paris 1977, 1269

[16] *K. V. Ettinger, N. A. Atari, J. R. Mallard*, Proc. of the Fifth Int. Conf. on Luminescence Dosimetry, Sao Paulo 1977, 40 (Ed. *A. Scharmann*)

[17] *K. V. Ettinger, R. W. Rowe, J. R. Mallard, A. Takavar, J. P. Sephon, D. I. Thwailes*, Proc. of the Fifth Int. Conf. on Luminescence Dosimetry, Sao Paulo 1977, 162 (Ed. *A. Scharmann*)
[18] *A. Kannan*, Int. J. Appl. Radiat. Isotopes **30** (1979) 258
[19] *V. V. Matjuškov, G. A. Michalčenko, V. Ja. Sigaev, J. V. Judin, A. V. Lymarev*, Isv. Akad. Nauk SSSR, Ser. Fiz. **41** (1977) 1364
[20] *K. J. Puite*, Proc. of the Fourth Int. Conf. on Radiat. Prot. Ass., Paris, 1977, 1273
[21] *K. J. Puite, D. L. J. M. Crebolder*, Phys. Med. Biol. **22** (1977) 1136
[22] *A. Takavar, K. V. Ettinger, J. R. Mallard, N. Atari, S. Srirath* Proc. of the Fifth Int. Conf. on Luminescence Dosimetry, Sao Paulo 1977, 169 (Ed. *A. Scharmann*)
[23] *Ja. A. Teteris*, Izv. Akad. Nauk Latv. SSR, Ser. Fiz., 1978, No. 3, 44
[24] *E. Wiedemann, G. C. Schmidt*, Ann. Physik und Chemie (Neue Folge) **54** (1895) 604

Originalarbeiten

Bestimmung des mittleren Energieaufwandes $\overline{W}_L$ zur Bildung eines Ionenpaares in Luft für ^{60}Co-γ-Strahlung

R. Kunze, O. Hecker

(Amt für Standardisierung, Meßwesen und Warenprüfung, Bereich Meßwesen, Fachabteilung Ionisierende Strahlung)[1])

Auf der Grundlage eines ionometrischen und eines kalorimetrischen Verfahrens zur Bestimmung der Energiedosis in Graphit ist im ASMW der $\overline{W}_L$-Wert für ^{60}Co-γ-Strahlung ermittelt worden. Für trockene Luft wurde folgender Wert erhalten:

$(\overline{W}_L/e) = (33{,}87 \pm 0.34)\ J/C$

Die Fehler des Verfahrens werden diskutiert. Das Ergebnis wird mit neueren veröffentlichten Werten verglichen.

Keywords

air; calorimetric dosemeters; cobalt 60; dosimetry; gamma radiation; ion pairs; ionization chambers; radiation doses

1. Einführung

Der Wert des mittleren Energieaufwandes zur Bildung eines Ionenpaares in Luft ($\overline{W}_L$-Wert) ist eine der wichtigsten Größen in der Dosimetrie ionisierender Strahlung, besonders im Hinblick auf die ionometrische Energiedosisbestimmung. In den letzten 20 Jahren wurden deshalb — vor allem mit der Vervollkommnung der dosimetrischen Meßtechnik — immer wieder Messungen zur Bestimmung des $\overline{W}_L$-Wertes durchgeführt [8].

Für Quanten- und Elektronenstrahlung geht die Ermittlung von $\overline{W}_L$ im allgemeinen auf das *Bragg-Gray*-Prinzip zurück. Danach gilt für die Energiedosisleistung $\dot{D}_M$ im — den luftgefüllten Hohlraum — umgebenden Material M die Beziehung:

$$\dot{D}_M = I_m \cdot s_L^M \cdot (\overline{W}_L/e)$$

bzw. $(\overline{W}_L/e) \cdot s_L^M = \dot{D}_M/I_m$

- I_m: auf die Masse des luftgefüllten Hohlraums bezogener Ionisationsstrom
- s_L^M: relatives, auf Luft bezogenes Massenbremsvermögen des Materials M
- $\overline{W}_L/e$: auf die Elementarladung bezogener Wert von $\overline{W}_L$.

Die Größe $\dot{D}_M$ wird gewöhnlich mit einem kalorimetrischen Verfahren bestimmt, wobei sich im wesentlichen drei Methoden herausgebildet haben, die auf einer

1. direkten Bestimmung der Energiedosisleistung $\dot{D}_M$ mit einem Dosiskalorimeter,
2. Bestimmung der Energieflußdichte mit einem Flußdichtekalorimeter oder
3. Bestimmung der Gesamtleistung der radioaktiven Strahlungsquelle mit einem totalabsorbierenden Kalorimeter

basieren.

Für die drei Verfahren ergeben sich damit nach der obigen Beziehung folgende Bestimmungsgleichungen[1]):

$$\frac{\overline{W}_L}{e} \cdot s_L^M = \frac{\dot{D}_M^K}{I_m} \cdot \prod_i K_i \quad (1)$$

$$\frac{\overline{W}_L}{e} \cdot s_L^M = \frac{\psi^K}{I_m} \cdot \left(\frac{\mu_a}{\varrho}\right)_M \cdot \prod_j K_j \quad (2)$$

$$\frac{\overline{W}_L}{e} \cdot s_L^M = \frac{N_{tot}^K}{4\pi R^2 I_m} \cdot \left(\frac{\mu_a}{\varrho}\right)_M \cdot \prod_k K_k \quad (3)$$

In den Beziehungen bedeuten:

- $\dot{D}_M^K$, ψ^K, N_{tot}^K: die kalorimetrisch bestimmten Größen Energiedosisleistung, Energieflußdichte bzw. Gesamtleistung einer Strahlungsquelle
- $(\mu_a/\varrho)_M$: Massenenergieabsorptionskoeffizient des Materials M
- $\prod_i K_i$, $\prod_j K_j$, $\prod_k K_k$: Produkt von Korrektionsfaktoren, die sich aus den einzelnen Meßverfahren ergeben
- R: Abstand Strahlungsquelle–Meßort bei der ionometrischen Messung.

[1]) Anschrift: DDR-1026 Berlin, Wallstr. 16

[1]) In den Beziehungen (1) bis (3) wurde s_L^M als unbekannte Größe mit auf der linken Seite belassen, weil alle diese Verfahren auch zur experimentellen Bestimmung von s_L^M eingesetzt werden können, wenn man $\overline{W}_L$ als gegeben voraussetzt.

In bezug auf ihre Genauigkeit sind die Verfahren (1) und (2) als gleichwertig zu betrachten. Die dritte Methode (3) hat den Vorteil, daß sie mit Strahlungsquellen kleiner Aktivität ($A \leqq 10$ GBq, gegenüber $A \geqq 10$ TBq bei den Verfahren (1) und (2)) realisiert werden kann, aber den Nachteil der größeren Meßunsicherheit. Mit dem Aufbau eines Energiedosiskalorimeters im ASMW ergab sich die Möglichkeit der Bestimmung von $\overline{W}_L$ für Quantenstrahlung nach (1).

2. Kalorimetrische Bestimmung der Energiedosisleistung

Mit dem in der Arbeit [1] beschriebenen quasiadiabatischen Differentialkalorimeter wurde die Energiedosisleistung $\dot{D}_C^K$ im Graphit nach der Beziehung

$$\dot{D}_C^K = F_{el} \cdot \Delta \dot{U}_a^s \cdot K_V \cdot K_W \cdot K_S \tag{4}$$

F_{el}: Empfindlichkeit des Kalorimeters, Gy/μV (elektr. Eichfaktor des Kalorimeters)

$\Delta\dot{U}_a^s$: Brückenverstimmung [μV] bei Zufuhr von Strahlungsenergie bezogen auf die Meßzeit t_M, s

K_V: Korrektionsfaktor für den Einfluß von Fremdmaterial im Absorber

K_W: Korrektionsfaktor für den Einfluß lokaler Wärmequellen

K_S: Korrektionsfaktor für den Einfluß von Sekundärstrahlung

bestimmt.

Als Strahlungsquelle diente eine ^{60}Co–γ-Strahlungsquelle mit einer Aktivität von ~ 30 TBq. Die Messungen erfolgten in einem kollimierten Strahlungsbündel ($d = 5$ cm) bei einem Quellenabstand von 185 cm. Beide Systeme des Differentialkalorimeters wurden wechselweise als Meßsysteme benutzt. Der Mittelwert aus beiden Systemen für die Energiedosisleistung $\dot{D}_C^K$ betrug $4{,}45 \cdot 10^{-2}$ Gy/min. Es wurden insgesamt 18 Einzelmessungen je System durchgeführt. In Tab. 1 sind die Werte der in die Bestimmungsgleichung (4) eingehenden Größen und die Fehler ihrer Bestimmung zusammengestellt. Die Meßunsicherheit der kalorimetrischen Messung, die nach [2] und [3] ermittelt wurde, betrug 0,8% bei $P = 95\%$.

Tab. 1. (Zur Bestimmungsgleichung (4))

Größe	Wert	nichterfaßter systematischer Fehler, %	
F_{el}[1])	$9{,}61 \cdot 10^{-2}$ Gy/μV $10{,}31 \cdot 10^{-2}$ Gy/μV	0,2	0,4[2])
t_M	20 min		
ΔU_a^s	8 ... 10 μV	0,2	0,4[2])
K_V	0,998	0,1	
K_W	1,005	0,2	
K_S	1,003	0,1	

[1]) für das rechte bzw. linke System
[2]) Gesamtfehler nach [3]

3. Ionometrische Messung

Mit dem Kalorimeter wurde eine Graphit-Hohlraumionisationskammer aufgebaut, die in ihren Abmessungen und im Material mit den Absorber-Jackettsystemen des Kalorimeters identisch ist. Bei dem gewählten Aufbau entspricht dem Absorber die Meßelektrode und dem Jackett die spannungsführende Elektrode. Damit die ionometrischen und kalorimetrischen Messungen unter gleichen äußeren Bedingungen erfolgen konnten, wurde weiterhin eine den äußeren Hüllen des Kalorimeters entsprechende Kalorimeterattrappe aufgebaut, in die die Kammer für die ionometrischen Messungen eingebracht wurde. Damit entfielen Korrektionen für die Schwächung und Streuung in den Kalorimeterhüllen.

Die Strommessung erfolgte mit einem Photozellenkompensator. Die Unterschiede der Strommessung in Abhängigkeit von der Polarität der Kammerspannung lagen innerhalb der Meßunsicherheit der Strommessung. Die Größe I_m wurde nach folgender Beziehung ermittelt:

$$I_m = \frac{I}{V \cdot \varrho_0} \cdot K_D \cdot K_F \cdot K_{St} \cdot K_M \tag{5}$$

I: Ionisationsstrom

V: Meßvolumen der Ionisationskammer

ϱ_0: Luftdichte ($p = 1{,}013 \cdot 10^5$ Pa; $T = 273{,}2$ K)

K_D, K_F: Korrektionsfaktor für die Luftdichte und Luftfeuchte

K_{St}: Korrektionsfaktor für die Stielstreuung

K_M: Korrektionsfaktor für die Meßortverschiebung gegenüber der kalorimetrischen Messung.

Tab. 2 enthält die Werte der in die Gleichung (5) eingehenden Größen sowie die Fehler ihrer Bestimmung.

Tab. 2. (Zur Bestimmungsgleichung (5))

Größe	Wert	nichterfaßter systematischer Fehler, %	
I	~ 10^{-10} A	0,2	0,3[1])
V	3,490 cm³	0,2	
K_D	1,08 ... 1,12	0,1	
K_F	1,002	0,1	
K_{St}	0,998	0,1	
K_M	0,994	0,2	

[1]) Gesamtfehler nach [3]

Aus insgesamt 35 Einzelmessungen für den Ionisationsstrom I über einen Zeitraum von einem halben Jahr ergab sich für I_m nach (5) ein Wert von $I_m = 2{,}170 \cdot 10^{-5}$ A/kg. Die Meßunsicherheit betrug 0,6% bei $P = 95\%$.

4. Bestimmung von $\overline{W}_L$ und Diskussion des Ergebnisses

Nach (1) erhält man mit den Beziehungen (4) und (5) für die korrigierten Werte von $\dot{D}_C^K$ und I_m folgende Bestimmungsgleichung für $(\overline{W}_L/e) \cdot s_L^C$:

$$(\overline{W}_L/e) \cdot s_L^C = \frac{\dot{D}_C^K}{I_m}$$

Mit den Werten $\dot{D}_C^K$ und I_m aus der kalorimetrischen und ionometrischen Messung erhält man damit für

$$(\overline{W}_L/e) \cdot s_L^C = 34{,}18 \text{ J/C}$$

Eine kritische Größe in diesem Verfahren zur Bestimmung von $\overline{W}_L$ ist das relative Massenbremsvermögen von Graphit zu Luft. Es wurde hier auf der Grundlage von [9] unter Berücksichtigung der Hohlraumgröße ein Wert von $s_L^C = (1{,}009 \pm 0{,}005)$ für ^{60}Co-γ-Strahlung erhalten. Damit ergibt sich für den mittleren Energieaufwand zur Bildung eines Ionenpaares in trockener Luft für ^{60}Co-γ-Strahlung der folgende Wert:

$$(\overline{W}_L/e) = (33{,}87 \pm 0{,}34) \text{ J/C}$$

Tab. 3. Werte für die W_L-Bestimmung

Autor/Jahr	Verfahren/Strahlungsquelle		$(\bar{W}_L/e)$, J/C
Engelke, Hohlfeld, 1971 [4]	1	^{137}Cs	33,72 ± 0,17¹)
Guiho, Simeon, 1975 [5]	1	^{60}Co	33,96 ± 0,34
Niatel, 1978 [6]	1	^{60}Co	33,72 ± 0,22
Engelke, Hohlfeld, 1970 [7]	2	^{137}Cs	34,1 ± 0,6
ICRU-Report Nr. 31 [8]	auf der Grundlage von Literaturanalysen ermittelter Wert		33,85 ± 0,15
diese Arbeit, 1979	1	^{60}Co	33,87 ± 0,34

¹) ohne Berücksichtigung des Fehlers für s_L^C

Die Meßunsicherheit setzt sich aus den statistischen Fehlern für die kalorimetrische und ionometrische Messung, aus den nichterfaßten systematischen Fehlern beider Meßverfahren und dem Fehler für s_L^C zusammen. Die Zusammenfassung der Einzelfehler erfolgte nach [3] für eine statistische Sicherheit von $P = 95\%$. In Tab. 3 sind Werte für die $\bar{W}_L$-Bestimmung aus den letzten 10 Jahren zusammengestellt.

Eingegangen am 14. 3. 1980

Literatur

[1] *R. Kunze, H. Rothe, H. Korn*, Isotopenpraxis **15** (1979) 265
[2] *S. Wagner*, PTB-Mitteilungen **79** (1969) 343
[3] ASWM-Richtlinie zur Berechnung der Meßunsicherheit (1978)
[4] *B. A. Engelke, K. Hohlfeld*, PTB-Mitteilungen **81** (1971) 336
[5] *J. P. Guiho, J. P. Simeon*, Int. J. Appl. Rad. Isot. **26** (1975) 714
[6] *M.-T. Niatel*, Rapport BIPM-77/8
[7] *B. A. Engelke, K. Hohlfeld*, PTB-Mitteilungen **80** (1970) 20
[8] ICRU-Report Nr. 31 (1979)
[9] *L. V. Spencer, F. H. Attix*, Rad. Res. **3** (1955) 239

Изменения нуклеиновых кислот в крови крыс после однократного рентгеновского облучения

Т. Сабова, Е. Мишурова

(Кафедра общей биологии факультета Естествознания УПЙШ, Кошице)¹)

Исследовались количественные изменения нуклеиновых кислот и инкорпорация ^3H-тимидина и ^{14}C-оротовой кислоты „ин виво“ (3 часа) в крови крыс в течение 7 суток после однократного облучения рентгеновскими лучами в дозе 154,8 мС/кг (600 Р). Наблюдалось резкое понижение концентрации ДНК до 12-ого часа после облучения, сопровождаемое понижением активности связанного трития в крови, однако удельная активность ДНК существенно не изменялась. В течение последующих дней концентрация ДНК оставалась на сниженном уровне. Инкорпорация ^3H-тимидина постепенно возрастала так, что на 7-ой день после облучения удельная активность ДНК была почти в 5 раз выше контрольных величин. Изменения РНК были мало выразительными и в большинстве случаев несигнификантными.

Keywords

blood cells; carbon 14 compounds; irradiation; nucleic acids; rats; tritium compounds; X radiation

Однократное и континуальное облучение животных вызывает изменения в составе крови, именно понижение количеств лейкоцитов, которое сопровождается изменением количества нуклеиновых кислот [1, 2].

В первых часах после однократного облучения в крови собак и морских свинок значительно понижается и количество клеток, синтетизирующих ДНК, но позже их количество возрастает, и на 10-ые и 40-ые сутки после облучения превосходит контрольные величины [3, 4]. Изменения состава крови, вызванные облучением, являются — кроме прямого повреждения элементов крови — в основном следствием торможения синтеза нуклеиновых кислот и митоза и морфологически наблюдаемых повреждений кроветворных органов, в результате чего нарушается образование новых клеток, необходимых для замены старых и поврежденных элементов крови.

Материал и методика

Опыты проводились на белых крысах-самцах линии Wistar. Вес животных перед началом опыта был около 180 граммов. Опытные животные подвергались однократному общему облучению экспозицей 154,8 *мС/кг* (600 *Р*) рентгеновских лучей и обследовались на 6, 12, 18, 24 и 30-ый час и на 2, 3 и 7-ые сутки. Облученным и контрольным животным 3 часа перед убоем вводились ^3H-тимидин и ^{14}C-оротовая кислота (55 мккюри ^3H и 4,7 мккюри ^{14}C на 100 *г* живого веса

¹) Адрес: УССР, 04167 Кошице, Шробарова 57

интраперитонеально), и был прекращен доступ к воде и пише.

Изменения нуклеиновых кислот определялись в гепаринизированной крови из аорты каждого животного в дупликатах. Эритроциты удалялись при помощи гемолизирующего раствора сапонина. Количество нуклеиновых кислот в седименте определялось по методу *Цанева* и *Маркова* [5]. Концентрация ДНК и РНК определялась на основании абсорбанции ультрафиолетовых лучей при двух длинах волн. Радиоактивность измерялась в жидком сцинтилляторе Брея на сцинтилляционной установке **LKB-Wallac** 81000. Полученные данные подвергались статистической обработке по методу анализа варианций и при помощи теста *Дункана* [6].

Результаты

В крови контрольных и облученных животных исследовались следующие показатели:

а) количество лейкоцитов
б) концентрация ДНК и РНК (*мг P*/100 *мл* крови)
в) удельная активность ДНК и РНК (*dpm*/*мг* ДНК или РНК)
г) активность связанного ^{3}H или ^{14}C в крови (*dpm*/*мл* крови).

Количество лейкоцитов резко понижалось в течение первых 12-и часов после облучения и на 7-ые сутки составляло только 10% от контрольных величин (табл. 1).

Суммарная концентрация нуклеиновых кислот (ДНК + РНК) после 6-ого часа понижалась почти в экспоненциальной зависимости от времени; временное повышение наблюдалось на 3-ие сутки после облучения.

Концентрация ДНК в периферической крови облученных животных резко понизилась от 6-ого до 12-ого часа. В последующие сроки концентрация ДНК оставалась на пониженном уровне в рамках 30—50% контролных величин (рис. 1).

Удельная активность ДНК (*dpm*/*мг* ДНК) до 72-ого часа достоверно не изменялась. Начиная с этого времени резко возрастала и на 7-ые сутки почти в 5 раз превосходила контрольные величины. Активность связанного в крови трития (*dpm*/*мл*) была в первые два дня временно понижена. В течение последующих дней активность постепенно возрастала к норме.

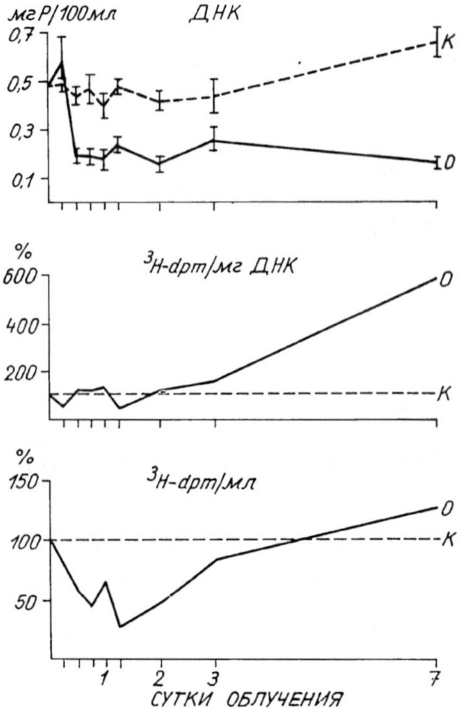

Рис. 1. Концентрация ДНК, удельная активность ДНК и активность связанного трития в крови крыс при облучении с мощностью дозы 154,8 *мС*/*кг* в течение 7 дней
К — контрольные, О — облученные животные

Изменения концентрации РНК, удельной активности РНК и активности связанного в крови ^{14}C были маловыразительными и в большинстве случаев несигнификантными. Статистически значимое различие между экспериментальными и соответствующими контрольными данными наблюдались только в концентрации РНК на 7-ые сутки после облучения (рис. 2).

Табл 1. Количество лейкоцитов и концентрация нуклеиновых кислот (ДНК ± РНК) в крови крыс после однократной экспозиции 154,8 мС/кг (К — контрольные, О — облученные животные)

Показатель		Интервал, часы							
		6	12	18	24	30	48	72	168
Лейкоциты количество/*мл*	К	8300 ±1609	12980 ±1231	8275 ± 998	9480 ±1237	9500 ±655	9233 ±1538	13625 ± 932	11840 ± 2477
	О	5875 ±1329	4000 ± 719	1860 ± 242	3533 ± 976	830 ± 131	1025 ± 217	2500 ± 851	1180 ± 311
Нуклеиновые кислоты (мг *P*/100 *мл*)	К	1,101 ±0,071	1,105 ±0,088	1,015 ±0,119	0,872 ±0,084	1,169 ±0,041	1,254 ±0,066	1,044 ±0,163	1,812 ±0,312
	О	1,083 ±0,215	0,845 ±0,089	0,676 ±0,168	0,690 ±0,138	0,788 ±0,086	0,659 ±0,115	1,143 ±0,216	0,839 ±0,080

В таблице приводится арифметическсе среднее ± ошибка среднего

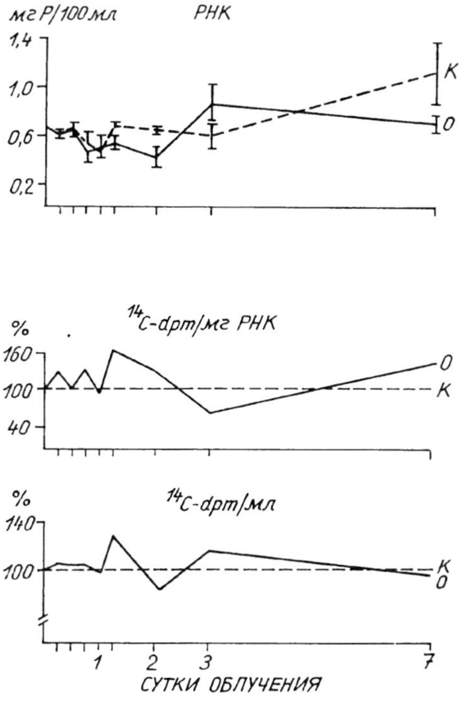

Рис. 2. Концентрация РНК, удельная активность РНК и активность связанной оротовой кислоты в крови крыс при облучении с мощностью дозы 154,8 *мС/кг* в течение 7 дней
К — контрольные, О — облученные животные

Дискуссия

Приведенные изменения суммарной концентрации нуклеиновых кислот (ДНК + РНК) согласуются с наблюдениями *Критского* и *Александрова* [1], которые после разных однократных облучений крыс (2—500 *Р*) определяли в течение первых двух суток экспоненциальное понижение концентрации суммарных нуклеиновых кислот с последующим частичным восстановлением. При континуальном облучении концентрация нуклеиновых кислот в крови понижается до 14-х суток подобным образом, в зависимости от продолжительности облучения [2]. Применение модифицированного метода [5] в этой работе показало, что количественные изменения РНК после облучения незначительны, и понижение концентрации нуклеиновых кислот в крови вызвано резким понижением концентрации ДНК.

Понижение концентрации нуклеиновых кислот в крови *Критский* и *Александров* [1] обьясняют распадом нуклеиновых кислот. Результаты нашего опыта показали, что понижение концентрации ДНК протекает в определенной мере одновременно со снижением количества лейкоцитов, но концентрация РНК существенно не изменяется. Таким образом, приведенные данные свидетельствуют о том, что после облучения происходит изменение и в наличии разных популяций и субпопуляций лейкоцитов с неодинаковым содержанием ДНК и РНК. Это предположение опирается на наблюдения *Едгрена* и *Вебера* [7], касающиеся очень радиорезистентной популяции лимфоцитов крови с повышенным содержанием РНК. Повышению величин концентрации нуклеиновых кислот в крови способствуют и другие элементы крови (в основном ретикулоциты) и плазма [8, 9], поэтому изменения концентрации этих веществ являются менее выразительными, чем изменения количества лейкоцитов.

Исследование инкорпорации радиоактивных предшественников после трехчасового мечения ,,ин виво‘‘ подтвердило выразительное влияние излучения на ДНК. При таком коротком времени мечения в периферическую кровь может поступить только небольшое количество новообразованных меченых клеток; изменения активности ДНК в крови считаем поэтому следствием изменения скорости инкорпорации предшественника в ДНК ядерных элементов крови, т. е. лейкоцитов.

Опыты с несколькочасовым мечением ,,ин витро‘‘ показали, что инкорпорация предшественников в нестимулированные лейкоциты происходит в основном при репликационном синтезе ДНК. В крови здоровых людей и животных количество клеток, синтетизирующих ДНК очень низкое, около 0,02—0,06% всех лейкоцитов [4, 10, 11]. При опеределенных патологических состояниях количество этих элементов в крови повышается [10, 12, 13].

Возрастание специфической активности ДНК в нашем опыте указывает на повышенное наличие ДНК синтетизирующих клеток в крови крыс на 7-ой день после облучения. Повышенное наличие клеток, синтетизирующих ДНК, в крови считаем одним из неспецифических проявлений, сопровождающих повышенную активность кроветворных и лимфатических тканей после повреждений. Это предположение обосновано на следующих наблюдениях:

1) Возрастание количества ДНК синтетизирующих клеток в циркуляции было определено при многих заболеваниях, например злокачественных опухолях [12, 13, 14], воспалительных процессах и после инфекции [15, 16].
2) После однократного облучения [4, 3 и наши результаты] и в течение континуального облучения [17] количество ДНК синтетизирующих клеток в крови повышается около 7-го—10-го дня, т. е. в период повышенной регенеративной активности кроветворных и лимфатических тканей.
3) Изменения количества ДНК синтетизирующих клеток в циркуляции при патологических состояниях не являются характеристическими только для определенного вида животных; до сих пор были определены у человека [12—14], собаки [4, 18], морской свинки [3] и крысы (наши результаты).

Большинство приведенных авторов отмечает, что ДНК синтетизирующие клетки относятся в основном к мононуклеарной серии. Их роль пока еще точно не известна и ее обьяснение требует дальнейших исследований.

Поступила в редакцию 10/III, 1980 г.

Литература

[1] *Г. А. Критский, С. В. Александров*, Доклады Академии наук СССР **209** (1973) 728
[2] *E. Mišúrová, T. Szabová*, Medicina lotnicza (in press)
[3] *P. F. Harris, G. Haigh, J. H. Kugler*, British Journal of Haematology **9** (1963) 385
[4] *V. P. Bond, F. M. Fliedner, E. P. Cronkite*, Journal of Laboratory and Clinical Medicine **57** (1961) 711
[5] *Р. Г. Цанев, Г. Г. Марков*, Биохимия **25** (1960) 151
[6] *D. E. Duncan*, Biometrics **11** (1955) 1
[7] *J. Edgren, T. A. Weber*, Acta Radiol. Therapy Physics Biology **15** (1976) 315

[8] *R. C. Kamm, A. G. Smith*, Clinical Chemistry **6** (1972) 519

[9] *R. A. Cox, M. Gokcen*, Biochemical Medicine 15 (1976) 126

[10] *S. W. A. Kuper, J. R. Bignall*, The Lancet **27** (1964) 1412

[11] *R. G. Evans, A. Norman*, Nature **217** (1968) 455

[12] *J. E. Sokal, J. L. Pauly*, Journal of Laboratory and Clinical Medicine 5 (1974) 681

[13] *H. E. Heier, T. Godal*, Scand. Journal of Haematology **18** (1977) 149

[14] *P. Schick, F. Trepel, H. Begemann*, Scand. Journal of Haematology **14** (1975) 17

[15] *V. P. Bond, T. M. Fliedner, E. P. Cronkite, J. R. Rubini, G. Brecher, P. K. Schork*, Acta Haematology **1** (1959) 1

[16] *D. A. Horwitz, P. Stastny, M. Ziff*, Journal Lab. Clinical Medicine **3** (1970) 391

[17] *T. Szaborá, E. Mišúrová*, Isotopenpraxis **16** (1980) 98

[18] *T. J. Yang, J. B. Dale, J. B. Jones*, Experimental and Molecular Pathology **25** (1976) 121

Modellrechnungen zur Isotopentrennung durch Multiphotonen-Photodissoziation von Molekülen

H. Johansen, A. Gräßer

(Zentralinstitut für Isotopen- und Strahlenforschung der Akademie der Wissenschaften der DDR)[1]

Für stoßfreie SF_6-Moleküle werden Multiphotonen-Photofragmentierungswahrscheinlichkeiten im intensiven IR-LASER-Feld berechnet einschließlich der Isotopenselektivität derartiger Prozesse in Abhängigkeit von der eingestrahlten LASER-Linie, der Leistungsdichte, dem Impulsprofil und der Temperatur. Über ein quantenmechanisches Modell werden die Rotationsschwingungsniveaus des SF_6-Moleküls, die Verstimmung zwischen LASER- und Molekülschwingungsfrequenz und die intramolekulare Schwingungsrelaxation berücksichtigt.

Keywords:

dissociation; infrared radiation; isotope effects; isotope separation; lasers; molecules; sulfur 32; sulfur 34

1. Einleitung

Auf der Grundlage eines quantenmechanischen Modells werden Multiphotonen-Photofragmentierungswahrscheinlichkeiten in intensiven LASER-Feldern für stoßfreie SF_6-Moleküle im elektronischen Grundzustand berechnet einschließlich der Isotopenselektivität derartiger Prozesse. Dabei werden die Fragmentierungswahrscheinlichkeiten in einem gepulsten CO_2-LASER-Feld in Abhängigkeit von der Impulsdauer, der Impulsform und der Leistungsdichte des LASERs ermittelt, wobei die Verstimmung zwischen der Frequenz der Photonen und der Oszillatorfrequenz einer geeigneten Normalschwingung des Moleküls berücksichtigt werden sowie die Anharmonizität der entsprechenden Schwingungsform, die intramolekulare Schwingungsrelaxation und schließlich die zu den Schwingungsniveaus der herausgegriffenen Schwingungsform gehörigen Rotationsniveaus und die Temperatur.

In Anlehnung an *Mukamel* und *Jortner* werden auf der Grundlage des Formalismus des effektiven *Hamilton*-operators zwei Modelle zur Berechnung der Fragmentierungswahrscheinlichkeiten herangezogen, das Modell des quasi-zweiatomigen Moleküls ohne Einbeziehung der Rotationsniveaus und das Zweileitermodell, hier als Modell der quasientarteten Zustände bezeichnet [2].

Nur wenn die Rechnungen am stoßfreien Molekül auf beträchtliche Isotopieeffekte ($10^2 - 10^3$) hinsichtlich der Fragmentierung hinweisen, kommen diese Moleküle für eine eventuelle Isotopentrennung in Frage, da in realen Systemen die Stöße zwischen den Molekülen den Isotopentrenneffekt deutlich herabsetzen.

[1]) Anschrift: 7050 Leipzig, Permoserstr. 15

2. Modell des quasi-zweiatomigen Moleküls ohne Einbeziehung der Rotationsniveaus

Das Modell des quasi-zweiatomigen Moleküls übernehmen wir von *Mukamel* und *Jortner*. Wir erwähnen hier nur die wichtigsten Eigenschaften des Modells und skizzieren den

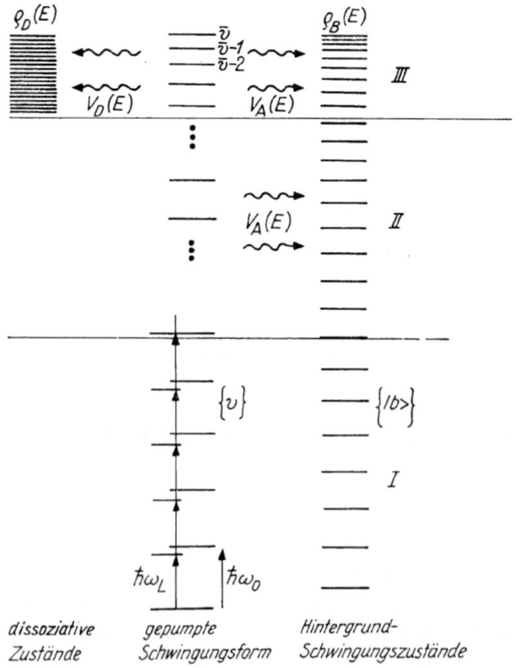

Abb. 1. Energieniveauschema eines mit der Frequenz ω_L optisch gepumpten Moleküls im Rahmen des Modells des quasizweiatomigen Moleküls

Weg der Berechnung der Fragmentierungswahrscheinlichkeiten. Ausführliche Darstellungen findet man in den Arbeiten [1–3]. Das Modell unterscheidet drei Energiebereiche innerhalb des Schwingungsenergieniveausystems des Moleküls. In Abb. 1 wird das Energieniveauschema eines mit der Frequenz ω_L optisch gepumpten Moleküls dargestellt. Die Frequenz ω_L befindet sich nahezu in Resonanz mit der Frequenz einer Normalschwingungsform des Moleküls (z.B. ν_3 im SF_6-Molekül). Die Schwingungsenergieniveauleiter der herausgegriffenen Schwingungsform wird durch die Quantenzahl v charakterisiert. Das Energieniveauschema des Hintergrundes wird durch die übrigen Normalschwingungsformen des Moleküls erzeugt (z.B. 14 Schwingungsformen im SF_6-Molekül.

Im Bereich I ist die Energieniveaudichte des Hintergrundes gering, jedoch im Bereich II sind die Niveaus (v) bereits quasientartet mit einer Mannigfaltigkeit von Schwingungszuständen des Hintergrundes. Die Kopplungsenergie $V_A(E)$ (E: erreichte Energie auf der Leiter; V_A: Energie der anharmonischen Kopplung zwischen den Schwingungsformen) markiert zusammen mit der Zustandsdichte des Hintergrundes $\varrho_B(E)$ über die Beziehung

$$V_A(E)\varrho_B(E) \gg 1 \tag{1}$$

den Anfang des Bereichs II. Im Bereich III öffnen sich dissoziative Kanäle.

Das Modell für die molekulare Multiphotonen-Photodissoziation beruht auf den folgenden Annahmen:

a) die Energieniveaus (v) werden als Niveaus eines anharmonischen Oszillators in der Näherung

$$E_v = \omega_0 v - [\omega_0 x]v^2;\; v = 0, 1, \ldots, \bar{v} \tag{2}$$

angesehen (ω_0: Oszillatorfrequenz; $[\omega_0 x]$: Anharmonizitätskonstante);

b) Rotationsschwingungsniveaus werden zunächst vernachlässigt (siehe Abschnitt 4);

c) den höheren Niveaus im Energiebereich II und den Niveaus in III werden Schwingungsrelaxationsbreiten Γ_v^{VR} zugeordnet;

d) das höchste Niveau ($v = \bar{v}$) wird durch eine Prädissoziationsbreite $\Gamma_{\bar{v}}^D$ charakterisiert, so daß die totale Breite des höchsten Niveaus $\Gamma_{\bar{v}} = \Gamma_{\bar{v}}^D + \Gamma_{\bar{v}}^{VR}$ beträgt;

e) während der Zeit $0 \leqq t \leqq t_f$ wirkt ein intensives elektromagnetisches Feld, beschrieben durch eine Zustandsfunktion $|n\rangle$, welche die Anhäufung von n Photonen mit der Frequenz ω_L in einer Schwingungsform des LASER-Systems beschreibt; die Verstimmung für den $0 \to 1$ Übergang auf der herausgegriffenen Energieniveauleiter beträgt

$$\Delta = \omega_0 - \omega_L$$

f) die Energieniveaus des Gesamtsystems ohne Berücksichtigung der Hintergrundzustände sind dann durch

$$E(v, n) = n\omega_L + v\omega_0 - [\omega_0 x]v^2 \tag{3}$$

gegeben;

g) es werden nur Matrixelemente der Wechselwirkung des elektromagnetischen Feldes mit dem (harmonischen) Oszillator zwischen aufeinanderfolgenden Zuständen berücksichtigt:

$$\langle v'n'|H_{int}|v\, n\rangle = \mu\mathcal{E}(t)\delta_{n', n\pm 1} \times \\ \times [(v+1)^{1/2}\delta_{v', v+1} + v^{1/2}\delta_{v', v-1}] \tag{4}$$

dabei sind μ das Dipolmatrixelement für den $0 \to 1$ Übergang und $\mathcal{E}(t)$ die Amplitude des elektromagnetischen Feldes; Die zeitliche Entwicklung des Systems von $\bar{v}$ Molekülzuständen in einem intensiven LASER-Feld $|0, n\rangle$, $|1, n-1\rangle, \ldots, |\bar{v}, n-\bar{v}\rangle$ – von denen einige mit den Hintergrundzuständen und die obersten mit einem dissoziativen Kontinuum gekoppelt sind – wird im Rahmen des Formalismus des effektiven *Hamilton*operators beschrieben [1–4].

Der Raum der Schwingungszustandsfunktionen des Moleküls wird dabei in zwei Unterräume zerlegt, und zwar in den Unterraum der Schwingungszustände (v) (Projektionsoperator P) und in den Unterraum der Zustände des Hintergrundes ($Q = I - P$). Der *Hamilton*operator des aus dem Molekül und dem elektromagnetischen Feld bestehenden Gesamtsystems, H, wird in der Form

$$H = H_0 + V \tag{5}$$

zerlegt, wobei V die Kopplung zwischen den Zuständen (v) und den Hintergrundzuständen beschreibt. Löst man aus der Zerlegung der Resolvente

$$(E - H)^{-1} = G(E) = PG(E)P + PG(E)Q + \\ + QG(E)P + QG(E)Q \tag{6}$$

den zum Unterraum (P) gehörigen Anteil heraus,

$$PG(E)P = P(E - PH_0P - PR(E)P)^{-1}P \tag{7}$$

so findet man den für die Beschreibung der zeitlichen Entwicklung der oben aufgeführten Molekülzustände $|v, n-v\rangle$ geeigneten effektiven Hamiltonoperator H_{eff} in der Darstellung

$$H_{eff} = P(H_0 + R(E))P \tag{8}$$

Der (level shift)-Operator R enthält die erwähnte Kopplung,

$$R(E) = V + VQ(E - QHQ)^{-1}QV \tag{9}$$

und führt Niveauverschiebungen $(\bar{\Delta})$ sowie Breiten (Γ) in den effektiven Hamiltonoperator ein,

$$R(E) = \bar{\Delta} - \frac{i}{2}\Gamma \tag{10}$$

wobei letztere auf die intramolekulare Schwingungsrelaxation und die Kopplung der obersten Niveaus an ein dissoziatives Kontinuum zurückzuführen sind.

Der effektive *Hamilton*operator nimmt in Matrixform in dem von den $\bar{v}$ Schwingungswellenfunktionen aufgespannten Unterraum des *Hilbert*raums die Gestalt an:

$$H_{eff} = \left\| \begin{matrix} \text{I} & \text{II} & \text{III} \\ E(0,n) & \mu\mathcal{E} & 0 & & \\ \mu\mathcal{E} & E(1,n-1) & \mu\mathcal{E}\sqrt{2} & & \\ & \ddots & \ddots & & \\ & \mu\mathcal{E}\sqrt{v} & E(v,n-v) - \frac{i}{2}\Gamma_v^{VR} & \mu\mathcal{E}\sqrt{v+1} & \\ & & \ddots & \ddots & \\ & & & \mu\mathcal{E}\sqrt{\bar{v}} & E(\bar{v},n-\bar{v}) - \frac{i}{2}\Gamma_{\bar{v}} \end{matrix} \right\| \tag{11}$$

und zwar für die Zeit $0 \leqq t \leqq t_f$. Der Anfangszustand des Gesamtsystems ist $|0, n\rangle$, und die Wahrscheinlichkeitsamplitude für die Besetzung des dissoziativen Zustands $|\bar{v}, n-\bar{v}\rangle$ zur Zeit t ist gleich

$$C_{\bar{v}}(t) = \langle 0, n|\, U(t, 0)\, |0, n\rangle = \\ - \frac{1}{2\pi i}\int e^{-iEt}\langle 0, n|\,(E - H_{eff})^{-1}\,|0, n\rangle\, dE \tag{12}$$

$U(t, 0)$ ist der Zeitevolutionsoperator. Die Amplitude $C_{\bar{v}}(t)$ stellt sich durch die komplexen Eigenwerte $\Lambda_j = \varepsilon_j - \frac{1}{2}i\gamma_j$ und die (nichtorthogonalen) Eigenvektoren $|j\rangle$ des effektiven *Hamilton*operators dar:

$$C_{\bar{v}}(t) = 0, \quad t < 0 \tag{13}$$

$$= \sum_j \langle v|j\rangle \langle 0|j\rangle \exp\left(-i\varepsilon_j t - \tfrac{1}{2}\gamma_j t\right), \quad 0 \leq t \leq t_f \tag{14}$$

$$= C_{\bar{v}}(t_f) \exp\left(-iE_{\bar{v}}t - \tfrac{1}{2}\Gamma_{\bar{v}}t\right), \quad t > t_f \tag{15}$$

und die Wahrscheinlichkeit für die Photofragmentierung $P_D(t)$ ist schließlich durch

$$P_D(t) = \Gamma_{\bar{v}}^D \int_0^t |C_{\bar{v}}(\tau)|^2 \, d\tau, \quad 0 \leq t \leq t_f \tag{16}$$

$$= P_D(t_f) + \exp\left[-\Gamma_{\bar{v}}^D(t - t_f)\right] |C_{\bar{v}}(t_f)|^2, \quad t > t_f \tag{17}$$

gegeben.

3. Berechnung von Isotopieeffekten im Rahmen des Modells des quasizweiatomigen Moleküls

Bis zur Erreichung der Molekülgrenze sind im Fall des SF_6-Moleküls etwa 50 Sprossen auf der Energieniveauleiter der herausgegriffenen Schwingungsform zu überwinden. Für die Gewinnung qualitativer Aussagen genügt es, mit 10 oder 15 Sprossen zu rechnen, zumal die benötigten Parameter wie z. B. die Größe der intramolekularen Schwingungsrelaxation nicht mit ausreichender Genauigkeit bekannt sind.

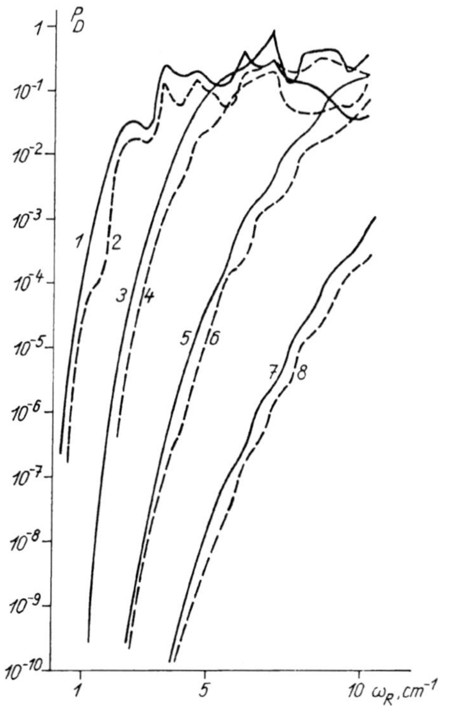

Abb. 2. Berechnete Fragmentierungswahrscheinlichkeiten für SF_6 in Abhängigkeit von der *Rabi*frequenz ω_R für zwei Impulsprofile [$^{32}\omega_0 = 948$ cm^{-1}; $^{34}\omega_0 = 931$ cm^{-1}; $|\omega_0 x| = 1$ cm^{-1}; $\bar{v} = 9$; $t_f = 167$ nsec; ①, ②: ^{32}S, $\Delta = 10$ cm^{-1}; ③, ④: ^{32}S, $\Delta = 0$ cm^{-1}; ⑤, ⑥: ^{34}S, $\Delta = 10$ cm^{-1}; ⑦, ⑧: ^{34}S, $\Delta = 0$ cm^{-1}; Rechteckimpuls: ①, ③, ⑤, ⑦; vom Rechteck abweichendes Profil: ②, ④, ⑥, ⑧ (siehe Text)].

Abb. 2 zeigt berechnete Fragmentierungswahrscheinlichkeiten von SF_6 im stoßfreien Zustand in Abhängigkeit von der *Rabi*frequenz ω_R für zwei Impulsprofile. Dabei erweist sich der Rechteckimpuls bei gleicher *Rabi*frequenz und Impulsdauer als effektiver, jedoch verliert sich diese Eigenschaft zum Teil im Sättigungsbereich. Der Isotopieeffekt nimmt bei gleicher Impulsdauer t_f für ein Impulsprofil der Gestalt

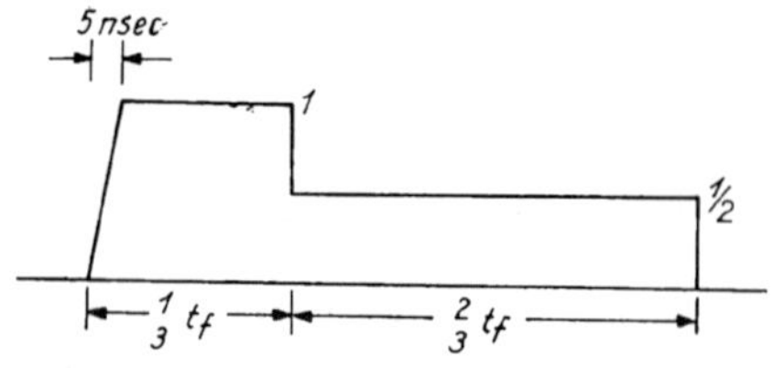

bis um den Faktor 10 zu.

Die berechneten Fragmentierungswahrscheinlichkeiten zeigen den bekannten Effekt, daß eine Verstimmung von der Art $\Delta = \bar{v}[\omega_0 x]$ für den Fragmentierungsprozeß am effektivsten ist (vgl. [1, 2] und Abb. 3). Isotopieeffekte von der Größenordnung $10^2 - 10^3$ im stoßfreien Zustand bei vertretbaren Fragmentierungsraten sind notwendig, um bei Drücken von 0,1 bis 1 torr Trennfaktoren in der Größenordnung von 10 zu erreichen.

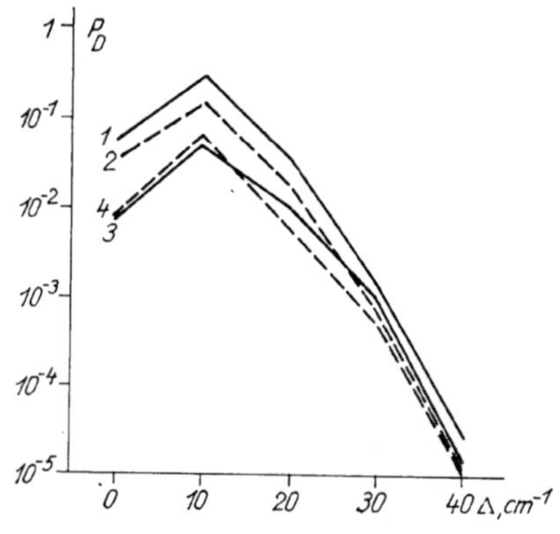

Abb. 3. Berechnete Frakmentierungswahrscheinlichkeiten für SF_6 bei fester *Rabi*frequenz in Abhängigkeit von der Verstimmung, der Impulsdauer (Rechteckimpuls) und der intramolekularen Schwingungsrelaxation

[$\omega_R = 10$ cm^{-1}; $|\omega_0 x| = 1$ cm^{-1}; $\bar{v} = 9$; ①, ②: $\Gamma_{\bar{v}}^D = 10^{-3}$ cm^{-1}, $\Gamma_v^{VR} = 0$ cm^{-1}, $v = 0, 1, \ldots 9$; ③, ④: $\Gamma_{\bar{v}}^D = 10^{-3}$ cm^{-1}, $\Gamma_9 = \Gamma_8 = 5 \times 10^{-3}$ cm^{-1}, $\Gamma_7 = \Gamma_6 = 10^{-3}$ cm^{-1}, $\Gamma_5 = 10^{-4}$ cm^{-1}, $\Gamma_4 = \ldots = \Gamma_0 = 0$ cm^{-1}; ①, ③: $t_f = 167$ nsec.; ②, ④: $t_f = 84$ nsec.]

Abb. 3 macht die Abhängigkeit der Fragmentierungswahrscheinlichkeiten von der Impulsdauer und der Verstimmung deutlich, ebenso die Dämpfung derselben als Folge der intramolekularen Schwingungsrelaxation. Die Energieniveaubreiten Γ_v für höher angeregte Schwingungsenergieniveaus sind schwer berechenbar und gehen daher nur als Parameter in die Rechnung ein.

4. Modell der quasientarteten Zustände (mit Einschluß der Rotationsschwingungsniveaus)

In diesem Modell werden die molekularen Energieniveaus, die nahezu in Resonanz mit der LASER-Energie $\hbar\omega_L$ liegen, in zwei Teilbereiche von Zuständen eingeteilt:

a) N_I Niveaus (v) im Bereich I (vgl. Abb. 4), wobei der effektive Hamiltonoperator im Teilraum der N_I Zustände (v) H_{eff}^A etwa die bisherige Gestalt, jedoch mit Einschluß der Rotationszustände annimmt (siehe unten);

b) N_{II} Niveaus in den Bereichen II, III, die zu quasientarteten Zuständen (gemischt aus Zuständen (v) und Hintergrundzuständen) gehören und in exakter Resonanz mit dem LASER-Feld liegen (H^{B}_{eff}).

Dem Modell liegt die Vorstellung zugrunde, daß das Molekül anfänglich mehrere Photonen isotopenselektiv absorbiert, d.h. für den Energiebereich I wird eine nahezu verwirklichte Resonanzkopplung zwischen dem LASER-Feld und den speziellen anharmonischen Zuständen (v) angenommen, wobei eine genügend hohe Rabifrequenz ω_R für die Überwindung der durch die Anharmonizität verursachten Niveauverschiebungen sorgt. Die immer größer werdenden Verschiebungen der Niveaus (v) in den Energiebereichen II, III verhindern jedoch die Absorption weiterer Photonen durch die Schwingungsform (v). Daher muß eine zweite Phase des Absorptionsprozesses wirksam werden, um das Molekül bis zur Bindungsdissoziationsenergie hin anzuregen. Das Modell nimmt im Gegensatz zum Modell des quasizweiatomigen Moleküls an, daß in den Bereichen II, III keine intramolekulare Schwingungsrelaxation und keine Umverteilung der Schwingungsenergie stattfinden, sondern vielmehr, daß die Dichte der gemischten Zustände genügend groß ist, um zu sichern, daß sich immer ein gemischter Zustand findet, der in exakter Resonanz mit dem Strahlungsfeld erreicht wird. Bei der Dichte der gemischten Zustände können Übergangsverbote außer acht gelassen werden. Das Modell enthält auch den Fall der nichtadiabatischen Kopplung verschiedener elektronischer Zustände.

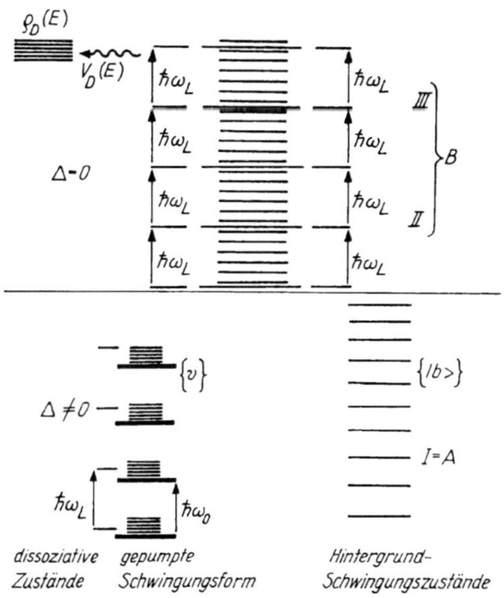

Abb. 4. Energieniveauschema eines mit der Frequenz ω_L optisch gepumpten Moleküls im Rahmen des Modells der quasientarteten Zustände

Der effektive *Hamilton*operator in Matrixform nimmt im Rahmen des Modells der quasientarteten Zustände näherungsweise die Gestalt an

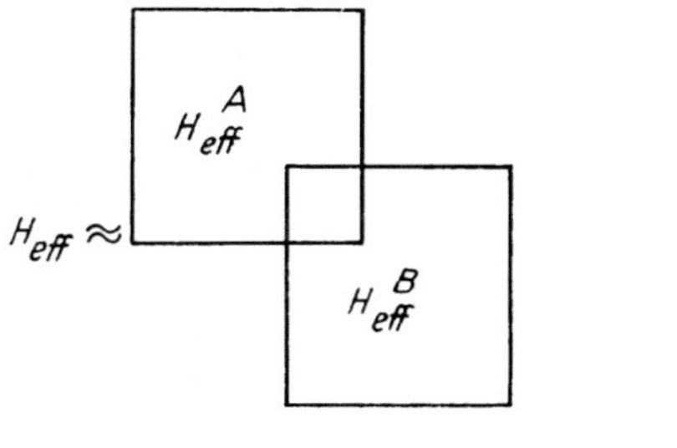

(18)

wobei sich die beiden Matrizen H^{A}_{eff}, H^{B}_{eff} in einem Diagonalelement überlappen.

Die Hauptdiagonalelemente des effektiven Hamiltonoperators im Teilraum der N_I anharmonischen Zustände (v) werden in der Form

$$\langle vI_v,\, n - v | H^{A}_{eff} | vI_v,\, n - v\rangle = E_{vib}(v) + B_v I_v(I_v + 1) - B_0 I_0(I_0 + 1) + (n - v)\omega_L - \frac{i}{2}\Gamma_v = (v, k_v) \tag{19}$$

$$E_{vib}(v) = \omega_0 v - [\omega_0 x] v^2 \tag{20}$$

$$I_v = I_0 + k_v, \quad k_v = -v, -v + 1, \dots, v \tag{21}$$

angesetzt. Dabei sind B_v, I_v ($v = 0, 1, \dots, N_I$) die Rotationskonstanten und molekularen Drehimpulseigenwerte in Abhängigkeit von der Schwingungsquantenzahl v; $[\omega_0 x]$ ist die Anharmonizitätskonstante. Die Nebendiagonalelemente werden aus dem Modell der effektiven Zustände von *Goodman*, das die Anharmonizität berücksichtigt, entwickelt [5, 6]:

$$\langle v,\, n - v | H^{A}_{int} | v + 1,\, n - (v + 1)\rangle = \delta_{n',\,n\pm 1} \frac{\mu \mathfrak{E} J(t)}{2} \left[\frac{(v + 1)(v + 2)(v + 3)}{2}\right]^{1/2} = (v,\, v + 1). \tag{22}$$

Dabei ist über Anfangs- und Endrotationszustände gemittelt worden. Der effektive Hamiltonoperator H^{A}_{eff} nimmt damit die Gestalt an

$$H^{A}_{eff} = \left\| \begin{matrix} (1,-1) & 0 & 0 & (1,2) & (1,2) & & \\ 0 & (1,0) & 0 & 0 & (1,2) & & \\ 0 & 0 & (1,+1) & 0 & 0 & & \\ (2,1) & 0 & 0 & (2,-2) & 0 & & \\ (2,1) & (2,1) & 0 & 0 & (2,-1) & & \\ & & & & & \ddots & \\ & & & & & & (v^*, k_{v^*} = v^*) \end{matrix} \right\| \tag{23}$$

wobei die Haupt- und Nebendiagonalelemente durch (19), (22) erklärt sind.

Aus der Quasiresonanzbedingung

$$E_{vib}(v^*) + B_{v^*}(I_0 + k_{v^*})(I_0 + k_{v^*} + 1) - B_0 I_0(I_0 + 1) - v^* \omega_L \approx 0 \tag{24}$$

wird eine Quantenzahl k_{v^*} zwischen $+v^*$ und $-v^*$ bestimmt und damit ein mit hoher Wahrscheinlichkeit erreichbares Rotationsschwingungsniveau, von dem aus aufeinander folgend N_{II} Niveaus im Energiebereich B in praktisch exakter Resonanz angeregt werden bis zur Molekülgrenze hin. Durch diese Auswahl des Zielniveaus im Energiebereich A wird eine entsprechende Wahrscheinlichkeitsamplitude für die Besetzung des Rotationsschwingungsniveaus (v^*, k_{v^*}) bestimmt, ${}^{A}C_{v=v^*}(t, I_0, k_{v^*}, \omega_L)$, wie unter (12), (14) beschrieben.

Die beiden effektiven *Hamilton*operatoren in Matrixform überlappen sich in dem Hauptdiagonalelement (v^*, k_{v^*}). Alle Hauptdiagonalelemente von H^{B}_{eff} weisen im Rahmen des Modells den gleichen Zahlenwert auf, nämlich denjenigen von (v^*, k_{v^*}). Die Bestimmung der Nebendiagonalelemente erfordert die Kenntnis der Wellenfunktionen der gemischten Zustände im Energiebereich B. Die Berechnung dieser Wellenfunktionen muß umgangen werden. Die Nebendiagonalelemente werden näherungsweise durch die mit einem Faktor $\varkappa < 1$ versehenen Nebendiagonalelemente $(v^* - 1, v^*) = (v^*, v^* - 1)$ aus dem Bereich A ersetzt:

$$(H_{\text{eff}}^{\text{B}})_{\text{v},\text{v}+1} \approx \varkappa(v^* - 1, v^*) \tag{25}$$

Der Parameter $\varkappa < 1$ berücksichtigt die verminderte Absorptionswahrscheinlichkeit der höheren Energiezustände.

$$H_{eff}^{B} = \left\| \begin{matrix} [v=v^*+1](v^*,k_{v^*}) & \varkappa(v^*-1,v^*) & & 0 \\ & [v=v^*+2](v^*,k_{v^*}) & & \\ & & \ddots & \\ & & & [v=\bar{v}](v^*,k_{v^*}) \end{matrix} \right\| \tag{26}$$

Mit Hilfe von $H_{\text{eff}}^{\text{B}}$ errechnet sich eine Wahrscheinlichkeitsamplitude für die Besetzung des obersten Niveaus $v = \bar{v}$

$${}^{\text{B}}C_{\text{v}=\bar{\text{v}}}(t, I_0, k_{\text{v}*}, \omega_{\text{L}})$$

Schließlich ergibt sich die Dissoziationswahrscheinlichkeit näherungsweise zu

$$P_{\text{D}}(t, I_0, k_{\text{v}*}, \omega_{\text{L}}) \approx \Gamma_{\bar{\text{v}}}^{\text{D}} \int_0^t |{}^{\text{A}}C(\tau)|^2 |{}^{\text{B}}C(\tau)|^2 \, \text{d}\tau, \quad 0 \leqq t \leqq t_{\text{f}} \tag{27}$$

Ausführlich geschrieben stellt sich P_{D} in der Abhängigkeit dar:

$$P_{\text{D}} = P_{\text{D}}(\omega_0, [\omega_0 x], \Gamma, I_0, v^*, k_{\text{v}*}, \omega_{\text{L}}, \omega_{\text{R}}, \varkappa, t_{\text{f}}, \tag{28}$$

Impulsprofil).

Eine Mittelung über die molekularen Anfangsdrehimpulseigenwerte mit den Gewichten $g(I_0)$ führt eine — wenn auch nicht umfassende — Temperaturabhängigkeit in die Berechnung der Dissoziationswahrscheinlichkeiten ein:

$$\bar{P}_{\text{D}}(\omega_{\text{L}}, T) = \frac{\Sigma_{I_0} g(I_0) e^{-\text{B}(0) \text{I}_0(\text{I}_0+1)/k_{\text{B}}\text{T}}}{\Sigma_{I_0} g(I_0) e^{-\text{B}(0) \text{I}_0(\text{I}_0+1)/k_{\text{B}}\text{T}}} P_{\text{D}}(I_0) \tag{29}$$

5. Berechnung von Isotopieeffekten im Rahmen des Modells der quasientarteten Zustände

Die Abb. 5 zeigt berechnete Fragmentierungswahrscheinlichkeiten $\bar{P}_{\text{D}}(\omega_{\text{L}}, T)$ von $^{32}SF_6$ und $^{34}SF_6$ für vier verschiedene LASER-Linien (Rechteckimpuls) in Abhängigkeit von der *Rabi*frequenz ω_{R}. Einer Leistungsdichte des LASERs von 1 GWatt/cm² entspricht im Fall des SF_6-Moleküls (Übergangsdipolmoment $\mu = 0{,}10$ D) eine *Rabi*frequenz von $\omega_{\text{R}} = 10 \text{ cm}^{-1}$. Die vier in Abb. 5 angegebenen LASER-Linien gehören zur 10,4 μm-Bande des CO_2-LASERs. Die gemittelten Fragmentierungswahrscheinlichkeiten $\bar{P}_{\text{D}}(\omega_{\text{L}}, T)$ sind abgeschätzt worden, indem nur Rechnungen für molekulare Drehimpulszustände in Abständen von $\Delta I_0 = 10$ im Intervall von $I_0 = 0$ bis $I_0 = 100$ durchgeführt worden sind und $v^* = 4$ gesetzt worden ist. Die Rotationskonstanten sind der Arbeit [5] entnommen worden. In fast allen Fällen liefert die Quasiresonanzbedingung (24) den Wert $k_{\text{v}*} = +4$.

Die Zahl der Energieniveaus im Bereich B ist mit $N_{\text{II}} = 20$ festgesetzt worden. Testrechnungen haben gezeigt, daß insbesondere der Isotopieeffekt in den Fragmentierungswahrscheinlichkeiten nur schwach von N_{II} abhängt. Der Parameter $\varkappa$ ist mit $\varkappa = 0{,}9$ angenommen worden. Ähnlich sind *Mukamel* und *Jortner* verfahren, die allerdings auch den Parameter $\varkappa < 1$ zufällig über die Nebendiagonalelemente variiert haben [2].

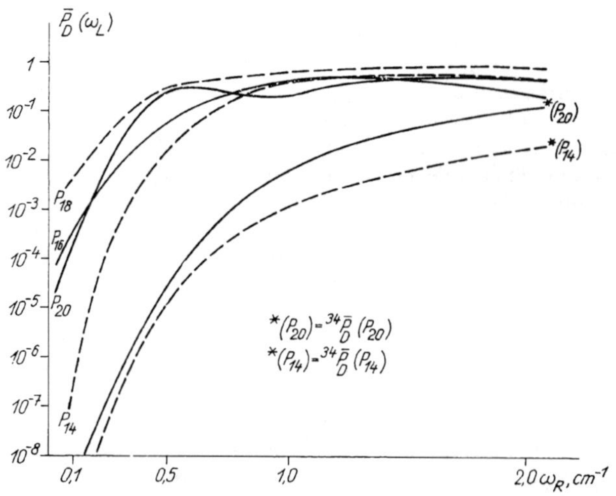

Abb. 5. Berechnete thermisch gemittelte Fragmentierungswahrscheinlichkeiten für SF_6 in Abhängigkeit von der *Rabi*frequenz ω_{R}. [$^{32}\omega_0 = 948 \text{ cm}^{-1}$; $^{34}\omega_0 = 931 \text{ cm}^{-1}$; $[\omega_0 x] = 2{,}56 \text{ cm}^{-1}$; $\omega_{\text{L}}(\text{P}(20)) = 944{,}18 \text{ cm}^{-1}$; $\omega_{\text{L}}(\text{P}(18)) = 945{,}9 \text{ cm}^{-1}$; $\omega_{\text{L}}(\text{P}(16)) = 947{,}7 \text{ cm}^{-1}$; $\omega_{\text{L}}(\text{P}(14)) = 949{,}43 \text{ cm}^{-1}$; Rechteckimpuls, $t_{\text{f}} = 167$ nsec, $T = 300$ °K].

Der glatte Verlauf der Fragmentierungswahrscheinlichkeiten $\bar{P}_{\text{D}}(\omega_{\text{L}}, T)$ in Abhängigkeit von der Rabifrequenz ω_{R} stellt eine Näherung dar, die auf einer Grobkörnigkeit bezüglich ω_{R} beruht. Die Dissoziationsbreite $\Gamma_{\bar{\text{v}}}^{\text{D}}$ ist $\Gamma_{\bar{\text{v}}}^{\text{D}} = 10^{-2} \text{ cm}^{-1}$ gesetzt worden.

Die berechneten Fragmentierungswahrscheinlichkeiten stehen in Übereinstimmung mit dem von Ackerhalt und Galbraith bis zum Niveau $v = 4$ berechneten Absorptionsspektrum, nach dem z.B. die P (16), P (18), P (20)-Linien eine höhere Absorptionswahrscheinlichkeit zeigen als die P (14)-Linie [5]. Wenn auch den berechneten Wahrscheinlichkeiten nur eine qualitative Bedeutung zukommt, so stehen doch die berechneten Isotopieeffekte im Einklang mit experimentellen Befunden am stoßfreien Molekül [7].

In Abb. 6 sind die Isotopieeffekte in den Fragmentierungswahrscheinlichkeiten für die vier LASER-Linien aufgetragen worden. Hohe Isotopieeffekte gehören zu niedrigen Fragmentierungsraten. Die inversen Temperatureffekte sind auf Sättigungseffekte oder auf ein Versagen des Modells für höhere *Rabi*frequenzen zurückzuführen.

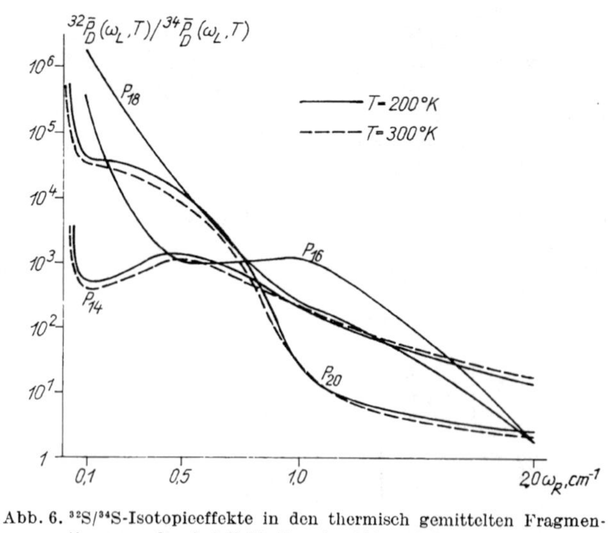

Abb. 6. ^{32}S/^{34}S-Isotopieeffekte in den thermisch gemittelten Fragmentierungswahrscheinlichkeiten in Abhängigkeit von der *Rabi*frequenz ω_{R} für verschiedene LASER-Linien und Temperaturen

Im Rahmen des Modells der quasientarteten Zustände werden die Abhängigkeit der Fragmentierungswahrscheinlichkeit $\overline{P}_D$ von der *Rabi*frequenz und damit von der Leistungsdichte des LASERs sowie die Isotopenselektivität des Fragmentierungsprozesses auf den untersten Energieniveaus (Bereich A) entschieden, da diese Größen nur schwach von der Anzahl N_{II} der aus dem Bereich B berücksichtigten Energieniveaus abhängen. Die Berücksichtigung der Rotationsniveaus gestattet es, die Abhängigkeit des Fragmentierungsprozesses von der eingestrahlten LASER-Linie adäquat zu behandeln.

Eingegangen am 10. 3. 1980

Literatur

[1] *S. Mukamel, J. Jortner*, Chem. Phys. Letters **40** (1976) 150

[2] *S. Mukamel, J. Jortner*, J. Chem. Phys. **65** (1976) 5204

[3] *J. Jortner, S. Mukamel*, in: The world of quantum chemistry, eds. *R. Daudel* and *B. Pullman* (Reidel, Dordrecht, 1973)

[4] *H. Johansen, K. Johst*. Proceedings of the Sixth international Conference on Atomic Physics, 1978, Riga

[5] *J. R. Ackerhalt, H. W. Galbraith*, J. Chem. Phys. **69** (1978) 1200

[6] *M. F. Goodman, J. Stone, D. A. Dows*, J. Chem. Phys. **65** (1976) 5052, 5062

[7] *J. L. Lyman, S. D. Rockwood, S. M. Freund*, J. Chem. Phys. **67** (1977) 4545

Zur Bestimmung der abnehmbaren Oberflächenkontamination

F. Schneider, H. Unger †

(Staatliches Amt für Atomsicherheit und Strahlenschutz, Außenstelle Lohmen)[1]

Es wird die Bestimmung der abnehmbaren Kontamination an Materialien, die in Kernanlagen und Radionuklidlaboratorien zum Einsatz kommen, behandelt. Vor allem wird auf das Problem des Abnahmefaktors beim Wischtestverfahren eingegangen. Die Größe des Abnahmefaktors hängt vom Zustand der kontaminierten Oberfläche, vom Zustand der Kontamination und von der Abnahmetechnik ab. In der Praxis erlaubt es die Unkenntnis der Auswirkungen von Einflußfaktoren auf den Abnahmefaktor oft nicht, diesen genau abzuschätzen. Auf der Grundlage der dargestellten Ergebnisse wird deshalb die Möglichkeit aufgezeigt, einen einheitlichen Abnahmefaktor von 0,5 einzuführen, ohne damit die bisher vorhandene Qualität des Strahlenschutzes zu mindern.

Keywords

decontamination; efficiency; radiation monitoring; radiation protection; surface contamination; surfaces

1. Einführung

Beim Vorliegen einer radioaktiven Oberflächenkontamination unterscheidet man zwischen einem „festhaftenden" (K_f) und einem „abnehmbaren" Anteil (K_a). Die Bestimmung von K_a erfolgt in der Regel[2]) durch Abwischen einer Oberfläche bekannter Größe mit einem geeigneten Material und durch anschließendes Ausmessen der Wischproben mittels eines Strahlungsmeßgerätes. Dieses Verfahren, in der Praxis als „Wischtest" bezeichnet, ist für den Strahlenschutz die wichtigste Methode zur Bestimmung der abnehmbaren Oberflächenkontamination, erlaubt es doch, Bestimmungen der Oberflächenkontamination auch an Orten durchzuführen, wo wegen zu hoher Umgebungsstrahlung keine Direktmessung möglich ist. Der beim Wischtest mit einmaligem Wischvorgpng entfernte Bruchteil von K_a wird als „Abnahmefaktor" (Q) bezeichnet[3]). Die Zusammenhänge zwischen Abnahmefaktor, abnehmbarer Kontamination und Gesamtkontamination (K_g) gehen aus Abb. 1 hervor. Gemäß der in der DDR gültigen „Richtlinie zur Bestimmung der Oberflächenkontamination" [1] ist der Abnahmefaktor in die Berechnung von K_a einzubeziehen:

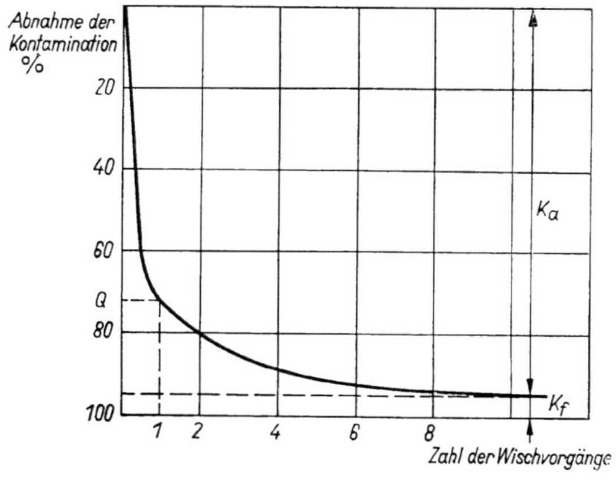

Abb. 1. Abnahme der Kontamination an PVC-Oberflächen mit der Zahl der Wischvorgänge
Kontaminant: ^{144}Ce-Lösung (eingetrocknet)

$$K_a = \frac{(n - n_0)\, f'}{S \cdot Q} \tag{1}$$

$(n - n_0)$ .... Nettoimpulsrate
f' .... Kalibrierungsfaktor der Meßanordnung
S .... abgewischte Fläche

Die Größe des Abnahmefaktors hängt von verschiedenen Einflüssen ab. Anhand eigener Meßergebnisse und anhand von Daten aus der Literatur wird nachfolgend ein Über-

[1]) Anschrift: 8345 Lohmen, Herrenleithe

[2]) Andere Bestimmungsmethoden bestehen in der Abnahme von Kontamination mittels Selbstklebeband oder im Absaugen mittels Luftstrom über Filter, wobei diese der Sammlung der abgesaugten Kontamination dienen.

[3]) Oft wird in der Literatur der Abnahmefaktor auch als Bruchteil der gesamten Oberflächenkontamination (K_g) aufgefaßt. Für dekontaminationsfreundliche Oberflächen, die in den folgenden Ausführungen vorwiegend in die Betrachtungen einbezogen werden, gilt jedoch $K_a \approx K_g$. Die Differenz zum Abnahmefaktor gemäß obiger Definition wird deshalb meist unerheblich sein.

blick gegeben, wie sich diese Einflüsse auf den Abnahmefaktor auswirken. In die Betrachtungen werden dabei nur solche Materialien einbezogen, die in Kernanlagen und Radionuklidlaboratorien Verwendung finden, die also einen geringen festhaftenden Kontaminationsanteil aufweisen und damit dekontaminationsfreundlich sind.

2. Einflüsse auf den Abnahmefaktor

2.1. *Zustand der kontaminierten Oberfläche*

Der Abnahmefaktor kann zuallererst durch die Art des vorliegenden kontaminierten Materials bzw. durch dessen Sorptionsverhalten gegenüber Radionukliden beeinflußt sein. Das Sorptionsverhalten hat insbesondere bei Vorliegen unverschmutzter Materialien Auswirkungen. In diesem Falle kann die Bindung der Kontaminationsnuklide direkt an der Materialoberfläche erfolgen, während im Falle des Auftretens von Schmutz- oder Korrosionsschichten ein Großteil durch diese gebunden sein wird. Die Abnahmefaktoren an unverschmutzten, verschmutzten und korrodierten Oberflächen sollten sich daher unterscheiden. Deutlich wird das am Beispiel des Verhaltens von Glas, das mit ^{137}Cs-Lösung kontaminiert ist, beim feuchten Wischtest [2]. Am sauberen Glas haftet das ^{137}Cs hauptsächlich auf Grund der Ionenaustauschfähigkeit der Glasoberfläche. Daher resultiert ein geringerer Abnahmefaktor als beim Vorliegen von Lehmverschmutzung, bei dem das ^{137}Cs offenbar an der leicht mechanisch entfernbaren Schmutzschicht gebunden ist. An anderen Werkstoffen, die in kerntechnischen Anlagen und Radionuklidlaboratorien häufig angewandt werden, Decelith H und nichtrostender Stahl, unterscheiden sich die Abnahmefaktoren zwischen unverschmutzten und öl- bzw. lehmverschmutzten Oberflächen dagegen kaum. Sie betragen beim feuchten Wischtest nach Kontamination mit Lösungen im Mittel etwa 0,90 [2]. Beim trockenen Wischtest nach Kontamination mit ThO_2-Staub liegen die Werte bei etwa 0,60 [3] (vgl. Tab. 1).

Der Einfluß der Oberflächenrauhigkeit auf den Abnahmefaktor bei Vorliegen trockener, staubförmiger Kontamination auf Plexiglas wurde von *Royster* und *Fish* [3] untersucht. Es kamen die schon eingangs genannten 3 verschiedenen Abnahmetechniken zur Anwendung (vgl. 1.).

Die Selbstklebetechnik zeigte eine proportionale Verringerung des Abnahmefaktors mit Zunahme der Rauhigkeit. Beim Absaugen der Kontamination mittels Luftstrom über ein Filter war dagegen kein Rauhigkeitseinfluß feststellbar. Für den Wischtest ergab sich im Bereich geringer Oberflächenrauhigkeit (Wurzel aus der mittleren quadratischen Abweichung $R < 5\,\mu m$) eine starke Verminderung des Abnahmefaktors mit Zunahme der Rauhigkeit.

Die Mehrzahl der in Kernanlagen und Radionuklidlaboratorien auftretenden Oberflächen wird jedoch höhere Werte von R ($R > 5\,\mu m$) aufweisen. Die von *Royster* und *Fish* [3] untersuchten Plexiglasoberflächen zeigten in diesem Bereich weitgehende Unabhängigkeit des Abnahmefaktors vom Rauhigkeitseinfluß. Faßt man den Wischtest als einen Dekontaminationsvorgang auf, so steht dieses Ergebnis aber im Widerspruch zu Erfahrungen aus der Dekontaminationspraxis. Aus diesen geht hervor, daß mit Zunahme der Rauhigkeit die Dekontaminierbarkeit abnimmt. Eine konkrete Aussage zu diesem Problem ist auf Grund des fast völligen Fehlens von experimentellen Daten bisher nicht möglich.

2.2. *Eigenschaften der Kontamination*

Von *Gorodinski* und *Goldstein* [4] wurden kationische Radionuklide hinsichtlich der Schwierigkeit ihrer Dekontaminierbarkeit in 4 Gruppen eingeteilt. Geht man wiederum davon aus, daß ein Wischtest letztendlich auch einen Dekontaminationsvorgang darstellt, so wäre mit dieser Gruppeneinteilung zu erwarten, daß die Abnahmefaktoren unterschiedlicher Radionuklide an jeweils gleichen Oberflächen verschieden groß sind. Untersuchungen zu diesem Problem sind von *Jakisch*, *Moldenhawer* und *Steuer* [2] vorgenommen worden. Es wurden verschiedene Materialien mit trägerfreien wäßrigen Lösungen der Radionuklide kontaminiert. Die Ergebnisse gibt Tab. 2 wieder.

Aus Tab. 2 geht hervor, daß die Art des Kontaminationsnuklids auf den Abnahmefaktor von Einfluß ist. Wegen der geringen Zahl an vorliegenden Ergebnissen kann jedoch nicht entschieden werden, ob die Gruppeneinteilung nach *Gorodinski* und *Goldstein* [4] auch für den Wischtest zutreffend ist.

Wegen der Vielfalt der in der Praxis möglichen Kontaminationsfälle hat man auch mit den unterschiedlichsten chemischen Formen des Auftretens einzelner Kontaminationsnuklide zu rechnen, was Einfluß auf den Abnahmefaktor haben kann. So besteht z.B. die Möglichkeit, daß ein Radionuklid in Form von Ionen, Molekülen, Kolloiden oder grobdispersen Partikeln an eine Oberfläche gebunden ist. Auch der Ladungszustand dieses Nuklids könnte unterschiedlich sein. Bei Vorliegen des Nuklids in grobdispersen Partikeln ist in Betracht zu ziehen, daß diese entweder oxidischer, hydroxidischer, salzartiger oder auch organi-

Tab. 1. Abnahmefaktoren an Oberflächen verschiedener Materialien bei unterschiedlichen Abnahmetechniken, Kontaminant ThO_2-Staub (nach *Royster* u. *Fish* [3])

Material	Selbstklebetechnik	Wischtest, trocken	Absaugen über Filter	
Polyäthylen	0,70	0,57	0,11	In Kernanlagen und Radionu-
Plexiglas	0,78	0,71	0,16	klidlaboratorien häufig
Formica	0,73	0,71	0,27	auftretende Materialober-
Glas	0,75	0,65	0,27	flächen
nichtrostender Stahl	0,68	0,51	0,11	
Preßspanholz	0,73	0,24	0,07	In Kernanlagen und Radio-
Asphalt	0,59	0,49	0,15	nuklidlaboratorien kaum
Beton, unbehandelt	0,56	0,40	0,22	auftretende Materialober-
Beton, ölig	0,44	0,38	0,01	flächen

Tab. 2. Abnahmefaktoren in Abhängigkeit von der Art des Kontaminationsnuklids an verschiedenen Materialien, feuchter Wischtest (nach *Jakisch* u. a. [2]

Kontaminationsnuklide	Materialien		
	Decelith H	Glas	nichtrostender Stahl
^{35}S	0,85		
^{36}Cl	0,95		
^{45}Ca	0,95	0,95	
^{60}Co	0,95		
^{89}Sr	0,95	0,95	
$^{90}Sr/Y$	0,65	0,95	0,90
^{91}Y	0,85		
^{137}Cs	0,85	0,50	0,90
^{144}Ce	0,65	0,95	0,50
^{210}Pb	0,80		
^{226}Ra	0,60		

scher Natur sein können. Schließlich kann auch die Größe von Kolloiden oder grobdispersen Partikeln wesentlichen Einfluß auf den Abnahmefaktor haben.

Tab. 3 zeigt zu Letztgenanntem Ergebnisse, die von *Fish* u. a. [5] durch Untersuchungen mit staubförmiger Kontamination erzielt wurden.

Bei vergleichsweise geringen Partikelgrößen ist der Abnahmefaktor kleiner als bei Vorliegen größerer Teilchen.

Tab. 3. Abnahmefaktoren von ThO_2-Staub an nichtrostendem Stahl in Abhängigkeit von der Partikelgröße bei unterschiedlichen Abnahmetechniken, trockener Wischtest (nach *Fish* u. a. [5])

Abnahmetechnik	Partikelgröße		
	Agglomerate bis zu 10 μ (Einzelpartikelgröße 0,01 – 0,1 μ)	1,5 μ Einzel-Partikel	5 μ Einzel-Partikel
Wischtest	0,49	0,96	0,86
Absaugen über Filter	0,10	0,58	0,75
Selbstklebetechnik	0,68	0,96	1,00

Der Effekt kann auf das zunehmende Wirksamwerden von Grenzflächenenergien mit zunehmendem Dispersionsgrad zurückgeführt werden, oder er ist durch das bessere mechanische Haften kleinerer Teilchen in den Zerklüftungen der Materialoberfläche bedingt.

Eine analoge Abhängigkeit des Abnahmefaktors vom Dispersionsgrad der kontaminierten Teilchen tritt auch auf, wenn die Oberflächenkontamination aus wäßrigen Lösungen heraus erfolgt. So ist im allgemeinen ein hoher Abnahmefaktor zu verzeichnen, falls die kontaminierenden Nuklide im grobdispersen Material von Salz- oder Oxidkrusten gebunden sind. Liegen dagegen hauptsächlich Ionen oder Kolloide vor, die die Kontamination verursachen, kann man eine vergleichsweise dazu geringere Abnahme durch den Wischtest erwarten. Ein solcher Fall tritt vor allem bei eingetrockneten, weitgehend trägerfreien Radionuklidlösungen auf. Ein Beispiel gibt Tab. 4 wieder. Hier zeigen die UO_3- und ^{144}Ce-Schmutz-Suspensionen sowie die Uranylazetatlösung, die beim Eintrocknen Krusten aus grobdispersen Teilchen bilden, weit höhere Abnahmefaktoren, als sie sich bei eingetrockneter, weitgehend trägerfreier ^{60}Co-Lösung ergeben.

2.3. *Abnahmetechnik*

Tab. 1, 3 und 4 enthalten auch Ergebnisse zum Wert des Abnahmefaktors bei den unterschiedlichen Abnahmetechniken. Daraus kann abgeleitet werden, daß mit feuchtem Wischtest in der Regel die höchsten Abnahmefaktoren zu erzielen sind. Insbesondere beim Vorliegen eingetrockneter, trägerarmer Lösungskontamination werden bei Anwendung der Selbstklebetechnik wesentlich geringere Abnahmefaktoren festgestellt als mit dem feuchten Wischtest (vgl. Tab. 4). Die kleinsten Werte ergeben sich, wie Untersuchungen mit staubförmiger Kontamination zeigten [3, 4] beim Absaugen über ein Filter (vgl. Tab. 1 u. 3).

2.3.1. Standardabweichungen der Ergebnisse verschiedener Abnahmetechniken. Durch *Royster* und *Fish* [3] sind die Standardabweichungen der nach den 3 verschiedenen Abnahmetechniken erhaltenen Meßwerte ermittelt worden. 21 Personen führten dazu Bestimmungen der abnehmbaren Oberflächenkontamination an Stahloberflächen von Arbeitsräumen aus. Jede der 21 Personen nahm 8 Proben je Abnahmetechnik. Die Ergebnisse der Tab. 5 zeigen, daß die größten individuellen Standardabweichungen und die größte Standardabweichung der Abnahmetechnik beim Wischtest auftreten. Offenbar ist das auf die Beeinflussung der Wischtestnahme durch personenabhängige Faktoren

Tab. 4. Einfluß von chemischen Formen der Kontamination auf den Abnahmefaktor bei unterschiedlichen Abnahmetechniken[1])

Chemische Form der Kontamination	feuchter Wischtest	Renafol	Heftpflaster	T-Band	Prenakor
Uranylazetlösung, eingetrocknet	0,58	0,31	0,50	0,21	0,24
UO_3-Suspension, eingetrocknet	0,60	0,82	0,76	0,52	0,64
^{144}Ce-Schmutzsuspension, eingetrocknet	0,77	0,60	0,77	0,54	0,48
^{60}Co-Lösungskontamination, eingetrocknet	0,40	0,17	0,11	0,10	0,08

[1]) Die aufgeführten Abnahmefaktoren sind Durchschnittswerte aus Einzelwerten, die an verschiedenen Materialien (V_2A-Stahl, C-Stahl, Aluminium, PVC (weich), Öllackanstrich, Chlorkautschukanstrich, Fichtenholz und Beton) erhalten wurden.

Tab. 5. Standardabweichungen bei verschiedenen Abnahmetechniken (nach *Royster* und *Fish* [3])

	Wischtest, trocken	Absaugen über Filter	Selbstklebetechnik
Standardabweichung der Abnahmetechnik	39%	11%	15%
mittlere individuelle Standardabweichung	17%	9%	13%

(Festlegung der abzuwischenden Fläche, Dauer des Wischens, Durchführung des Wischtestes u. a.) zurückzuführen.

Die angegebenen Standardabweichungen der Abnahmetechnik können als Maße für die Genauigkeit der Abnahmefaktoren an Stahloberflächen angesehen werden.

2.3.2. Einfluß des Wischmaterials. Untersuchungen von *Dinter* [6] ergaben bei Verwendung verschiedener Wischmaterialien (Luftfilterblättchen, Baumwollpopeline, Zellwollpopeline, Leinen) keine wesentlichen Unterschiede im Abnahmefaktor. An Luftfilterblättchen stört jedoch vor allem an rauhen Oberflächen die relativ geringe Abriebfestigkeit. Der Papierabrieb kann bei der Bestimmung der Oberflächenkontamination zu deren Unterbewertung führen. Er tritt in stärkerem Maße beim feuchten als beim trockenen Wischtest auf.

Falls die Oberflächenkontamination durch α-strahlende Nuklide verursacht wird, ergibt sich die Notwendigkeit der Wischtestnahme nicht mehr aus der Beeinflussung der Direktmessung durch die Umgebungsstrahlung. Vielmehr sind hier Wischteste erforderlich, wenn Oberflächenbereiche durch α-Kontaminationssonden nicht zugänglich sind.

Wegen der geringen Reichweite der α-Strahlen sind spezielle Wischmaterialien, die eine wenig poröse Oberfläche aufweisen (z.B. Transparentpapier, Polyäthylenfolie), einzusetzen. Untersuchungen zu Abnahmefaktoren dieser Wischmaterialien sind bisher noch nicht publiziert worden.

2.3.3. Einfluß der Befeuchtung des Wischmaterials. Einen Vergleich der Abnahmefaktoren von feuchter und trockener Wischtestnahme gibt Tab. 6 wieder.

Die Ergebnisse der Tabelle weisen aus, daß mit trockenem Wischtest nur etwa die Hälfte der Kontamination entfernt wird, die man mit feuchtem Wischtest abnimmt. Weiterhin zeigen diese Ergebnisse, daß es kaum von Einfluß auf den Abnahmefaktor ist, wenn man anstelle von Wasser als Befeuchtungsmittel Dekontaminationslösung verwendet. Im Verlaufe des zeitmäßig sehr kurzen Wischvorganges von 6 bis 10 s kommen dekontaminationsfördernde Zusätze zum Wasser offenbar wenig zur Wirkung.

Tab. 6. Einfluß des Befeuchtungsmittels und der Art des Wischens auf den Abnahmefaktor an Öllackanstrich, kontaminiert mit ^{60}Co-Lösung

Zustand des Wischmaterials	Art des Wischens		
	a	b	c
trocken	0,37	0,41	0,16
angefeuchtet mit Wasser	0,91	0,88	0,82
angefeuchtet mit Dekontaminationslösung	0,90	0,86	0,70

a – Wischmaterial 6mal mit leichtem Andruck auf kontaminierter Stelle drehen
b – Wischmaterial 3mal kreuzweise über kontaminierte Stelle führen
c – Wischmaterial 1mal kreuzweise über kontaminierte Stelle führen

2.3.4. Einflüsse mit geringen Auswirkungen. Der Auflagedruck beim Wischtest hat im Bereich von 15 bis 60 kPa sowohl an glatten als auch an rauhen Oberflächen keine Auswirkungen auf den Abnahmefaktor [6]. Die dieser Aussage zugrunde liegenden Untersuchungen wurden mit der trockenen Wischtestmethode durchgeführt. Auflagedrucke > 60 kPa sind in der Praxis kaum zu erwarten, da es schon einer gewissen Kraftanstrengung bedarf, einen Druck von 60 kPa zu erreichen.

Um den Einfluß der Art des Wischens auf den Abnahmefaktor zu ermitteln, wurden 3 verschiedene Wischmethoden angewandt (vgl. Tab. 6). Die Ergebnisse der Tab. 6 zeigen bei gleichem Zeitaufwand für das Wischen (Wischarten a und b jeweils 6 s) keinen Unterschied im Abnahmefaktor. Wird jedoch die Wischart auf 3 s verkürzt (Wischart c), so ergeben sich niedrigere Abnahmefaktoren. Daraus kann verallgemeinernd geschlußfolgert werden, daß die Art des Wischens selbst wenig Einfluß auf den Abnahmefaktor ausübt, daß aber die Wischdauer sich deutlich auf diesen auswirkt.

Untersuchungen zum Einfluß der Größe der abgewischten Fläche auf den Abnahmefaktor sind nur von staubförmiger Kontamination her bekannt [3]. Im Flächenbereich < 70 cm² nimmt mit zunehmender Flächengröße der Abnahmefaktor kontinuierlich ab. Im Flächenbereich > 70 cm² wirkt sich eine weitere Zunahme der abgewischten Fläche nur noch wenig verringernd auf den Abnahmefaktor aus (Abb. 2).

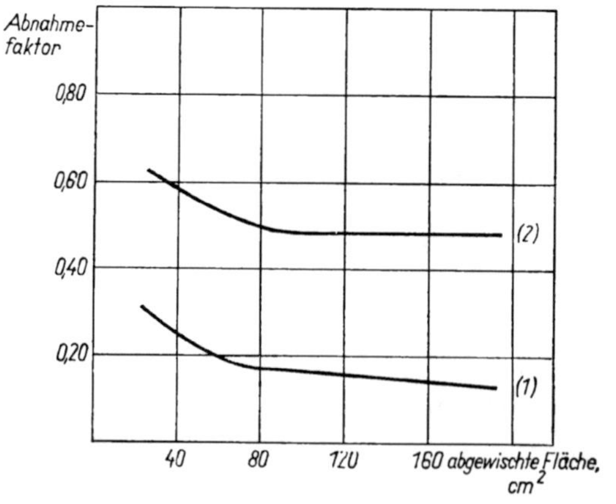

Abb. 2. Abnahmefaktor als Funktion der Größe der abgewischten Fläche bei Kontamination mit ThO_2-Staub, trockener Wischtest nach (*Royster* u. *Fish* [3])
1 – Preßspanholz, gehobelt; 2 – Formica

3. Schlußfolgerungen

In der praktischen Anwendung des Wischtestes ist man meist auf grobe Schätzwerte des Abnahmefaktors angewiesen, denn nur selten liegt für einen konkreten Kontaminationsfall auch ein experimentell bestimmter Wert vor.

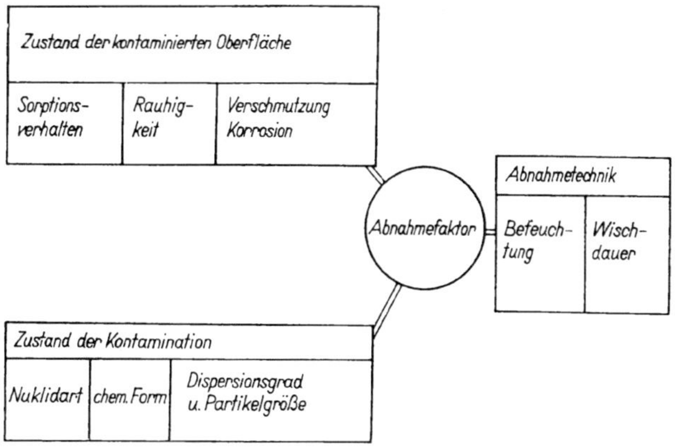

Abb. 3. Einflüsse mit starken Auswirkungen auf den Abnahmefaktor (von geringem Einfluß sind: Art des Wischmaterials und des Befeuchtungsmittels, Art des Wischens, Auflagedruck, Größe der abgewischten Fläche)

Ausgangspunkte für die in der Praxis erfolgende Abschätzung des Abnahmefaktors sind im wesentlichen die Oberflächenrauhigkeit und die Art des kontaminierten Materials. Die anderen der in Abb. 3 noch einmal zusammengestellten Einflüsse finden dagegen kaum Berücksichtigung. Die Schwierigkeiten in der richtigen Abschätzung sollten deshalb Anlaß dazu sein, das Problem des Abnahmefaktors im Rahmen zukünftiger Festlegungen zur Bestimmung der Oberflächenkontamination erneut zu erörtern.

Es wird empfohlen, für alle Materialien, die in Kernanlagen und Radionuklidlaboratorien zum Einsatz kommen, einen gleich großen Abnahmefaktor anzusetzen. In diesem Falle wäre bei feuchtem Wischtest ein Wert von 0,5 als ausreichend dafür anzusehen, die bisher vorhandene Qualität des Strahlenschutzes beizuhalten, denn die meisten der in Kernanlagen und Radionuklidlaboratorien auftretenden Oberflächen zeigen Abnahmefaktoren $> 0,5$ (vgl. Tab. 2, 4, 6). Die beim trockenen Wischtest gegenüber dem feuchten Verfahren etwa um den Faktor 0,5 verminderte Abnahme (vgl. Tab. 6) müßte durch Einbeziehung des Faktors „2" ind die Gleichung zur Berechnung der abnehmbaren Oberflächenkontamination (1) Berücksichtigung finden.

Eingegangen am 20. 3. 1980

Literatur

[1] „Richtlinie zur Bestimmung der Oberflächenkontamination" vom 4. Juni 1974 (Mitteilungen des Staatlichen Amtes für Atomsicherheit und Strahlenschutz 1974 Nr. 4)

[2] *N. Jakisch, F. Moldenhawer, J. Steuer*, Richtlinie zur Bestimmung der Oberflächenkontamination SZS-7/68

[3] *G. W. Royster, B. R. Fish*, Techniques for Assessing "Removable" Surface Contamination, in: Surface Contamination, Pergamon Press 1967, S. 201–207

[4] *S. M. Gorodinski, D. S. Goldstein*, ZBNTI' Moskau 1967

[5] *B. R. Fish, R. L. Walker, G. W. Royster*, Jr., *J. L. Thompson*, Redispersion of Settled Particulates, in: Surface Contamination, Pergamon Press 1967, S. 75–81

[6] *H. Dinter*, Untersuchungen zum Wischtestverfahren, Praktikumsbericht KKW Rheinsberg, Abt. Dosimetrie 1967

Kurze Originalmitteilungen

Über die Bestrahlung von Getreidestroh mit energiereicher Strahlung (1. Mitt.)

M. Bär, J. Leonhardt

(Zentralinstitut für Isotopen- und Strahlenforschung der AdW der DDR)[1]

G. Flachowsky, A. Hennig, I. Wolf

(Sektion Tierproduktion und Veterinärmedizin der Karl-Marx-Universität Leipzig, Wissenschaftsbereich Tierernährungschemie)[2]

K. Nehring

(em. Professor für Tierernährung)[3]

Keywords

animal feeds; biological radiation effects; digestibility; gamma radiation; sheep; straw

In der Welt fallen jährlich ca. 2 Milliarden t Getreidestroh an, die nur partiell für die Fütterung genutzt werden, weil seine Verdaulichkeit gering ist. Durch verschiedene chemische und physikalische Methoden ist seit den neunziger Jahren des vorigen Jahrhunderts versucht worden, die Verdaulichkeit des Strohs zu erhöhen (Übersicht *Flachowsky* [1]). 1962 berichteten *Pritchard* u. a. [2] über den Aufschlußeffekt von Gamma-Strahlen bei Stroh. In vitro konnten sie die Verdaulichkeit beträchtlich erhöhen, in vivo erzielten *Sandev* und *Karaivanov* [3] bei allerdings geringerer Dosis keinen Effekt auf die Verdaulichkeit der organischen Substanz. Dieser Widerspruch bewog uns, den Effekt der Gamma-Bestrahlung des Strohs sowohl in vivo als in vitro vergleichend zu prüfen. In unseren Versuchen wurden Weizenstrohproben in der Kobalt-60-Hochdosisleistungs-Gammabestrahlungsanlage GUP 32 PANORAMA des Zentralinstitutes für Isotopen- und Strahlenforschung der AdW, Leipzig, bestrahlt. Mit dieser Anlage ist es möglich, die Dosisleistung in weiten Grenzen bis max. 9 kGy/h zu variieren. Auf die Weizenstrohproben wurden Gesamtdosen zwischen 100 Gy und 3 MGy appliziert. Die Be-

Anschriften: [1]) 7050 Leipzig, Permoserstr. 15; [2]) 6900 Jena, Dornburger Str. 24; [3]) 2500 Rostock, Schliemannstr. 15a

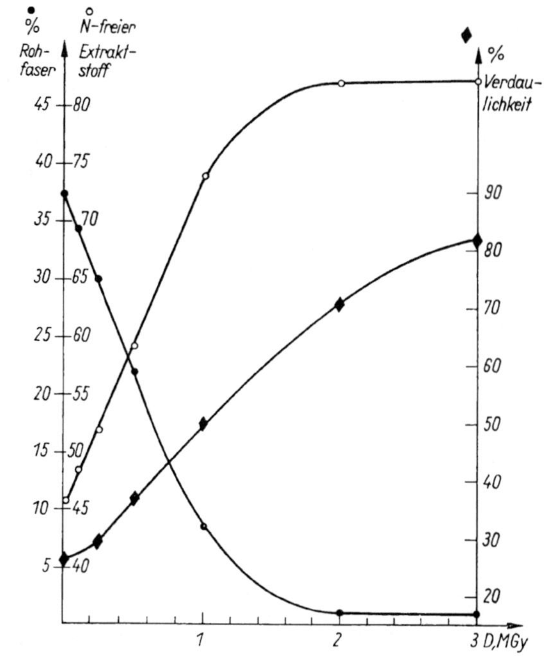

Abb. 1. Rohfasergehalt, Verdaulichkeit und N-freier Extraktstoff von bestrahltem Weizenstroh

stimmung der Verdaulichkeit in vivo erfolgte in Versuchen mit Schafen und in vitro mit der Zellulasemethode (*Nehring* [4]).

Durch die Anwendung energiereicher ionisierender Strahlung sinkt mit zunehmender Dosis ab 0,25 MGy der Rohfasergehalt deutlich ab (Abb. 1). Die kleinen Bruchstücke in bestrahlteden n Proben entgehen der Rohfaserbestimmung. Die Verdaulichkeit in vitro wird durch die Bestrahlung von ≈ 20 auf 80% erhöht. In den Verdaulichkeitsversuchen mit Schafen konnte sowohl mit Langstroh als auch Strohpellets das Ergebnis der in-vitro-Versuche bestätigt werden. Die Verdaulichkeit der organischen Substanz des Strohs erhöhte sich um 10 bzw. 15 Einheiten. Es sind daher vergleichbare Effekte wie beim Aufschluß des Strohs mit NaOH erzielbar, wobei die optimale Bestrahlungsdosis in Tierversuchen noch ermittelt werden muß. Über die Ergebnisse der einzelnen Befunde wird demnächst ausführlich berichtet.

Eingegangen am 8. 4. 1980

Literatur

[1] *G. Flachowsky*, Einsatz von Strohpellets in der Fütterung. VEB Deutscher Landwirtschaftsverlag Berlin, 1976

[2] *G. I. Pritchard, W. J. Pigden, D. J. Minson*, Canad. J. Animal Sci. **42** (1962) 215

[3] *S. Sander, I. Karaivanov*, Tierernähr. Fütt. **10** (1977) 238

[4] *K. Nehring*, 13. Internationaler Graslandkongreß, 18.–27. 05. 1977 in Leipzig, Sektion 10. Sektionsvorträge 475 (1977)

Informationen

IAEA. Die Internationale Atomenergie-Agentur (IAEA) hat in einem Pressekommunique (PR 80/8) „Einige Bemerkungen zu Strahlungen und geringen Strahlendosen sowie zur Besorgnis der Öffentlichkeit bezüglich ihrer Wirkungen" publiziert, die von einer Expertengruppe der IAEA erarbeitet wurden. Eine etwas gekürzte Fassung dieses Pressekommuniques wird nachfolgend wiedergegeben.

„In Anbetracht der Bedenken, die in letzter Zeit in gewissen Kreisen bezüglich der genetischen und anderen Wirkungen schwacher Strahlungen laut geworden sind, und um unbegründete Befürchtungen der Öffentlichkeit zu zerstreuen, hält die IAEA folgende Ausführungen für angebracht:

Das Problem der schwachen Strahlungen wird – in einigen Fällen seit mehr als 50 Jahren – von zahlreichen internationalen Expertengruppen studiert, die dazu Empfehlungen ausgearbeitet haben.

1. Im Jahre 1928 wurde die Internationale Strahlenschutzkommission (ICRP = International Commission on Radiological Protection), eine unabhängige nichtstaatliche Expertengruppe, gegründet, um Empfehlungen darüber auszuarbeiten, welchen maximalen Strahlendosen Menschen ohne Gefährdung ausgesetzt werden können. Ihre Mitglieder werden auf Grund ihrer Qualifikation auf den Gebieten Röntgenologie, Strahlenschutz, Physik, Strahlenschutzphysik, Biologie, Genetik, Biochemie und Biophysik ausgewählt. Die Empfehlungen der ICRP werden sowohl auf innerstaatlicher Ebene als auch international von den für den Strahlenschutz zuständigen Stellen allgemein anerkannt.

2. Im Jahre 1955 setzte die Generalversammlung der Vereinten Nationen in Anbetracht der allgemeinen Besorgnis hinsichtlich der Wirkungen des radioaktiven Niederschlags infolge der Kernwaffenversuche eine weitere internationale Expertengruppe ein, den Wissenschaftlichen Ausschuß der UN über die Wirkungen atomarer Strahlen (UNSCEAR = United Nations Scientific Committee on the Effects of Atomic Radiation). Er wurde beauftragt, Daten über die in der Umwelt beobachteten Intensitäten der natürlichen und der künstlich hervorgerufenen ionisierenden Strahlungen und der Radioaktivität sowie über deren Wirkungen auf den Menschen und seine Umwelt zu sammeln, zu studieren und zu verbreiten. Der jüngste Bericht des Ausschusses wurde 1977 veröffentlicht.

3. Seit 1957 gibt die IAEA in Zusammenarbeit mit der Weltgesundheitsorganisation (WHO) und anderen Gremien Grundnormen für den Strahlenschutz heraus, die als Grundlage für die Gesetzgebung der einzelnen Staaten dienen.

Über die biologischen Wirkungen von Strahlungen auf den Menschen weiß man wesentlich besser Bescheid als über jene der meisten anderen Schadstoffe, die in die Umwelt gelangen, da seit Jahrzehnten in großem Umfang Daten gesammelt, analysiert und weiterverfolgt werden, so z. B. über die Wirkungen von Bestrahlungen zur Krebstherapie oder zur Diagnostik und Behandlung von Schilddrüsenerkrankungen sowie über jene von Röntgenbestrahlungen. Darüber hinaus wurde umfangreiches Material durch Tierversuche gewonnen.

Um die relative Wirkung geringer Strahlendosen zu illustrieren, hat das UNSCEAR in seinem Bericht des Jahres 1977 die Wirkungen der Belastung durch verschiedene Strahlenquellen mit jenen der natürlichen Hintergrundstrahlung verglichen, d. h. der Strahlung, der jeder Mensch jeden Tag unweigerlich ausgesetzt ist.

Diese natürliche Strahlendosis schwankt von Ort zu Ort. In New York oder Paris beträgt sie z.B. ca. 100 mrem/Jahr, in Denver (Colorado) 125 mrem und in Kerala (Indien) 400 mrem. Vergleichende Untersuchungen über die Wirkungen verschiedener Strahlendosen auf ähnliche Bevölkerungsgruppen haben jedoch keine schlüssigen Ergebnisse gebracht. Dies u. a. deshalb, weil es sehr schwierig ist, bei großen Bevölkerungsgruppen geringfügige Veränderungen im Auftreten von Krankheiten zu beobachten, wie sie durch geringe Strahlenbelastungen hervorgerufen werden könnten, und sie eindeutig von ähnlichen Erscheinungen zu unterscheiden, die durch eine Fülle anderer Ursachen bedingt sein könnten, z.B. wirtschaftliche Faktoren oder mangelhafte Ernährung. Veränderungen der Erkrankungshäufigkeit können auch einfach deshalb beobachtet werden, weil infolge einer besseren ärztlichen und sozialen Betreuung bessere diagnostische Einrichtungen zur Verfügung stehen.

Man kann nicht genug betonen, daß Entwicklungsschäden auch auf natürliche Weise auftreten, und daß es, um einen gewissen Anstieg ihrer natürlichen Häufigkeit auf einen bestimmten äußeren Einfluß zurückzuführen, jahrelanger, sorgfältig geplanter epidemiologischer Untersuchungen bedarf. Wegen der natürlichen statistischen Schwankungen der Häufigkeit solcher Schäden ist das eine gewaltige Aufgabe.

Je geringer die Strahlendosis ist, desto weniger Wirkungen treten auf. Deshalb muß man, um einen schlüssigen Beweis dafür zu erbringen, daß bestimmte gesundheitliche Wirkungen mit einer bestimmten minimalen Strahlendosis zusammenhängen, um so mehr Menschen beobachten, je geringer die Strahlendosis ist. So wurde z.B. berechnet, daß man, um eine Änderung der Häufigkeit genetischer Schäden nachzuweisen, die durch eine Erhöhung der Strahlendosis um 3 mrem pro Person und Jahr verursacht sein könnte, 700 Millionen Menschen drei Generationen hindurch beobachten müßte.

In einer Studie über die gesundheitlichen Wirkungen der Kernenergiegewinnung ist die WHO u. a. zu dem Schluß gelangt, daß ‚die durchschnittliche Strahlenbelastung der Bevölkerung durch die Gewinnung von Kernenergie lokal und global betrachtet selbst bei höherer Intensität gering ist im Vergleich zur duchschnittlichen Belastung durch natürliche Strahlenquellen oder medizinische Maßnahmen‘ (siehe: Health Implications of Nuclear Power Production, WHO Regional Publications, European Series Nr. 3, Kopenhagen 1978).

In Großbritannien wurde berechnet, daß von der gesamten Strahlenbelastung des Menschen mehr als 30% auf medizinische Bestrahlungen (Röntgenstrahlung) und 67% auf die natürliche Hintergrundstrahlung zurückzuführen sind.

Um das Problem in die richtige Perspektive zu stellen, hat das UNSCEAR berechnet, daß die jährliche Strahlenbelastung durch sämtliche derzeit in der Welt bestehenden Kernkraftwerke der Belastung durch natürliche Strahlung während 0,6 Tagen entspricht. Diese Berechnung zeigt, daß die von Kernkraftwerken freigesetzte Strahlung bisher keinerlei biologisch nachweisbare Wirkungen beim Menschen gehabt hat. Dabei beruhen alle diese hypothetischen Feststellungen auf mathematischen Extrapolationen (wobei eine lineare Dosis-Effekt-Beziehung angenommen wird). Die bei diesen Berechnungen verwendeten äußerst geringen Dosiswerte haben nur dank der Empfindlichkeit und Genauigkeit der modernen Strahlennachweisgeräte Bedeutung gewonnen. Die Messung der biologischen Wirkungen dieser geringen Dosen auf den Menschen ist aber um viele Größenordnungen weniger genau als die Dosismessung, selbst wenn die empfindlichsten diagnostischen Verfahren angewendet werden."

BRD. Im Kernforschungszentrum Karlsruhe wurde ein Verfahren zur Naßverbrennung plutoniumhaltiger Rückstände entwickelt, dessen Produkte in eine gut endlagerungsfähige Form übergeführt werden können. Die Hauptverfahrensschritte bestehen in der Verkohlung der zerkleinerten organischen Abfälle mit Schwefelsäure bei wenig erhöhter Temperatur und der anschließenden Oxydierung des zurückbleibenden Kohlenstoffs mit Salpetersäure zu Kohlendioxid. Als Rückstände fallen die mineralischen Bestandteile der Abfälle und das in eine lösliche Form gebrachte Plutonium an, das zu über 90% zurückgewonnen werden kann. Das Ausgangsvolumen der Abfälle wird auf rund 5% verringert. Mit der errichteten Versuchsanlage soll nun im halbtechnischen Maßstab die Zuverlässigkeit des Verfahrens im Dauerbetrieb geprüft werden.

Großbritannien. In Windscale wird in Kürze die Errichtung einer Pilotanlage zur Verglasung hochaktiver flüssiger Rückstände abgeschlossen werden. Seitens der British Nuclear Fuels Ltd. (BNFL) wurde mitgeteilt, daß die Laborergebnisse bei der Verglasung hochaktiver Rückstände hinsichtlich der Verfahrenstechnik und der Glasqualität sehr gut ausgefallen sind. Befürchtete unerwünschte Nebeneffekte durch Radiolyse und Gasbildung seien nicht aufgetreten. Die Pilotanlage wird pro Liter Flüssigabfall ca. 2,5 kg Glas liefern.

Es wird nun an gleicher Stelle mit dem Bau einer Demonstrationsanlage im industriellen Maßstab begonnen, deren Betriebsbeginn für 1987 geplant ist. Diese nach dem Harvest-Verfahren arbeitende Anlage soll 3 Durchlaufstraßen erhalten und auch die bis dahin angesammelten hochaktiven flüssigen Rückstände verarbeiten. Jede Verarbeitungslinie könnte pro Jahr 30–50 Glasblöcke mit einem Durchmesser von rund 46 cm und einer Länge von 2 m liefern, die zur Endlagerung in Edelstahlbehälter eingeschlossen werden sollen.

Ein weiteres Projekt auf dem Gelände der Wiederaufarbeitungsanlage von Windscale betrifft die Errichtung eines Komplexes für die Entwicklung und Prüfung von Komponenten für den Verglasungsprozeß (u. a. Verglasungsöfen und Vorrichtungen für die Materialzufuhr).

USA. Das in der Nähe des Argonne Nat. Lab. gelegene Laboratorium in New Brunswick hat u. a. folgende Aufgaben: Anfertigung zuverlässiger Analysen von Kernmaterialien auf Anforderung staatlicher Einrichtungen; Sammlung, Verarbeitung und Verbreitung von Informationen auf dem Gebiet der Analyse von Kernmaterialien; Ausarbeitung und Verbesserung von Analysenmethoden für Kernmaterialien; Beurteilung der Qualität der Durchführung von Analysen von Kernmaterialien in anderen

Laboratorien; Vermittlung der Analysentechnologie für Kernmaterialien und technische Unterstützung von staatlichen und privaten Organisationen.

Das Laboratorium ist in der Lage, jährlich bis zu 1500 Proben hinsichtlich ihres Urangehalts zu analysieren. Die Mitarbeiter des Laboratoriums entwickelten und eichten ein System zur automatischen Urangehaltsbestimmung in Proben mittels gravimetrischer Titration, das sich jetzt im Routinebetrieb befindet. Dieses System ermöglicht Messungen mit einem Fehler von weniger als 0,01% bei einem Urangehalt von 15 ... 150 mg. Die Zahl der durchführbaren Analysen konnte auf das Doppelte erhöht werden.

IAEA. Nummer 1/1980 des Bulletins der Internationalen Atomenergie-Agentur enthält u. a. folgende Beiträge:
Report of the International Consultative Group on Nuclear Energy;
Judgement in Achieving Protection Against Radiation (*L. S. Taylor*);
The IAEA International Uranium Geology Information System (*L. Trocki*);
Current Nuclear Power Plant Safety Issues: Preview of an International Conference (*M. Rosen*, *R. Schmidt*);
How Safe is "Too" Safe? (*S. C. Black*, *F. Niehaus*);
Safety Principles Underlying the NUSS Documents (*D. G. Hurst*).

IAEA. Neue Publikationen der Internationalen Atomenergie-Agentur:
Radiological Surveillance of Airborne Contaminants in the Working Environment (Safety Series No. 49 – STI/PUB/484);
Uranium Deposits in Africa: Geology and Exploration (Panel Proc. Series – STI/PUB/509);
Status and Prospects of Thermal Breeders and their Effect on Fuel Utilization (Tech. Rep. Series No. 195 – STI/DOC/10/195).

IAEA/Österreich. Am 28. April 1980 billigte der Gouverneursrat der Internationalen Atomenergie-Agentur (IAEA) ein Kontrollabkommen zwischen der Agentur und der Republik Kuba auf Grund eines zwischen der Sowjetunion und Kuba abgeschlossenen Abkommens über Zusammenarbeit bei der Errichtung eines Kernkraftwerks.

Das sowjetisch-kubanische Abkommen sieht unter anderem wirtschaftlich-technische Zusammenarbeit beim Bau eines Kernkraftwerks mit einer Kapazität von 880 MWe in Kuba und bei der Beschaffung der erforderlichen Ausrüstungen, Geräte und Materialien vor.

IAEA/Frankreich. Alljährlich werden auf der ganzen Welt schätzungsweise 8 Millionen Sendungen mit radioaktivem Inhalt per Flugzeug, zu Land und zur See befördert. Ihre Radioaktivität ist unterschiedlich, und bei ihrem Inhalt kann es sich um Uranerz, Leuchtzifferblätter, Rauchdetektoren, Radionuklide für medizinische Zwecke, radioaktive Rückstände oder bestrahlten Kernbrennstoff handeln. Praktisch sämtliche Sendungen dieser Art müssen den Sicherheitsbestimmungen für die Beförderung radioaktiver Stoffe (Schriftenreihe Strahlenschutz, Nr. 6, revidierte Ausgabe 1973) der Internationalen Atomenergie-Agentur (IAEA) entsprechen. Diese Vorschriften sollen beim Transport von Sendungen radioaktiven Inhalts mit konventionellen Transportmitteln auf den Verkehrswegen der Welt den Schutz der Öffentlichkeit und des Transportpersonals gewährleisten.

Die Ständige Beratungsgruppe der IAEA für die sichere Beförderung radioaktiver Stoffe (SAGSTRAM) tagte vor kurzem in Paris, um Vorschläge für die bevorstehende Revision der Beförderungsvorschriften und für das Programm der IAEA auf dem Gebiet der Sicherheit bei Transporten im allgemeinen auszuarbeiten. Fachleute aus 15 Mitgliedsstaaten der IAEA und Vertreter von 7 internationalen Organisationen, die sich mit der Beförderung potentiell gefährlicher Stoffe befassen (EG-Kommission, Kernenergieagentur, Weltpostverein, Internatianale Organisation für Standardisierung, Zwischenstaatliche Beratende Seeschiffahrtsorganisation, Weltgesundheitsorganisation, Zentralbüro für Internationale Eisenbahntransporte) waren beteiligt.

Da beim Transport radioaktiver Stoffe bereits ein sehr hoher Grad an Sicherheit erreicht wurde, empfahl die Gruppe, eine Änderung der Vorschriften nur dann in Betracht zu ziehen, wenn erwiesen ist, daß der Gewinn an Sicherheit größer wäre als der voraussichtliche Kostenaufwand. In den revidierten Vorschriften, deren Veröffentlichung die IAEA für 1983 plant, sollen alle Angaben in den Einheiten des neuen Internationalen Systems ausgedrückt werden, um sie an andere internationale Normen anzugleichen.

Die Gruppe forderte die IAEA auf, sich stärker für die Anwendung und Einhaltung der Vorschriften einzusetzen, um das bisherige hohe Sicherheitsniveau bei Transporten zu halten. Sie hob ferner hervor, daß Angaben über Sendungen radioaktiver Stoffe und über Unfälle bei Transporten gesammelt werden müßten, und empfahl der IAEA, Untersuchungen über die Einschätzung des Transportrisikos durchzuführen. Auf Wunsch der Zwischenstaatlichen Beratenden Seeschiffahrtsorganisation setzte man sich besonders mit den Anweisungen auseinander, die im nächsten Jahr den Kapitänen von Schiffen erteilt werden sollen, die radioaktive Substanzen transportieren.

Großbritannien. Die Fa. The Radiochemical Centre, Amersham, hat ihre Exportpreisliste 1980 (gültig ab April 1980) für umschlossene Strahlenquellen (Industrieanwendungen, Laboratoriumszwecke) herausgegeben. Die Preisliste enthält Angaben über einige neue Erzeugnisse und veränderte Code-Nummern gegenüber dem Katalog 1977/78 und der Ergänzung von 1978.

UdSSR. Heft 4/1980 der Zeitschrift „Atomnaja Energija" enthält u. a. folgende Arbeiten:
Eine radiochemische Schnellmethode zur Bestimmung der Dampffeuchtigkeit bei KKW mit Siedewasserreaktoren (*L. N. Moskwin* u. a.);
Vorhersage der technologischen Eigenschaften von Uranerzen in natürlichen Vorkommen mit Hilfe eines Regressionsmodells (*A. A. Derjagin* u. a.);

Kurze Originalmitteilungen:
Effektivität des Nachweises von Spaltprodukten mit einem Festkörper-Spurdetektor auf der Grundlage von Glimmer (*B. M. Alexandrow* u. a.);

Orientierungsänderungen des elektrischen Widerstands und der inneren Reibung von Niob-Einkristallen bei der Einwirkung von Temperatur und Strahlung (*M. I. Sacharowa* u. a.);
Optimales Profil der Abschirmung gegen Streustrahlung (*B. K. Bylkin* u. a.);
Berechnung des Gammaspektrums in einer Unterwasser-Kobaltbestrahlungsanlage (*I. R. Entinson*);
Über die Möglichkeit der Bildung von Säuren beim Betrieb von Kernreaktoren mit CO_2-Wärmeträger (*M. A. Kurbanow* u. a.);
Einfluß der Art der Heliumimplantation auf das Blistering von rostfreiem Stahl (*W. W. Gann* u. a.).

BRD. Im Kernforschungszentrum Karlsruhe wird an der Entwicklung eines Verfahrens zum Einschluß von ^{85}Kr im Kristallgitter von Zeolithen gearbeitet. Ein solches Verfahren könnte eine Alternative zur Endlagerung des bei der Wiederaufarbeitung von bestrahlten Kernbrennstoffen anfallenden ^{85}Kr in Druckflaschen darstellen.

Es wurde festgestellt, daß die Aufnahmekapazität von sog. 5A-Zeolithen (Aluminosilikate) hinsichtlich der erforderlichen Volumina mit denen der Druckflaschenspeicherung vergleichbar ist. Die zu den Gerüstsilikaten gehörenden 5A-Zeolithe besitzen im Inneren des Kristallgitters relativ große Hohlräume, die durch Öffnungen mit einem Durchmesser von 2×10^{-7} bis 4×10^{-7} mm miteinander verbunden sind. Beim Erhitzen auf 500 °C werden die Öffnungen so weit vergrößert, daß die Hohlräume unter Anwendung eines Drucks von einigen 100 bar mit Krypton gefüllt werden können. Nach Abkühlung auf Umgebungstemperatur ist das Krypton zuverlässig eingeschlossen. Auch bei längerer Temperaturbelastung bis auf 450 °C wurden keine nennenswerten Mengen an Krypton freigesetzt.

BRD. Im Auftrag der Transnuklear GmbH entwickelte und baute die NUKEM GmbH eine fahrbare Anlage zur Verfestigung radioaktiv kontaminierter Abfälle aus kerntechnischen Anlagen (MOWA), mit der vor Ort kontaminierte Schlämme, Konzentrate, Kugelharze und getrocknete Rückstände mit einer spezifischen Aktivität bis zu 1000 Ci/m^3 nach dem Zementierungsverfahren konditioniert werden können. Die Rückstände werden direkt im Abfallfaß mit Zuschlagstoffen und Zement vermischt.

Schweiz. Im Eidgenössischen Institut für Reaktorforschung wurden die physikalischen Parameter von Kugeln aus metallischem ^{237}Np mit einer Dichte von 20,45 g/cm^3 bestimmt. Für solche Gebilde beträgt der kritische Radius 8,86 cm; die kritische Masse 59,7 kg. Die Halbwertzeit von ^{237}Np für Alphazerfall beträgt $2{,}14 \times 10^6$ a, für spontane Spaltung über 10^{18} a. Es entsteht in Kernreaktoren durch Neutroneneinfang aus ^{235}U unter Bildung von ^{237}U und anschließender Umwandlung in ^{237}Np. In einem üblichen Leichtwasserreaktor mit einer elektrischen Leistung von 1 GW können pro Jahr bis zu 13 kg ^{237}Np erzeugt werden. Eine mögliche Weiterverwendung von ^{237}Np besteht in gasgekühlten schnellen Reaktoren mit U-Pu-Np-Brennstoffzyklus. Wie Berechnungen ergaben, beträgt der jährliche Bedarf eines solchen Reaktors an ^{237}Np ungefähr 536 kg/GWe, wobei 35% davon gespalten werden und der Rest in ^{238}Pu umgewandelt wird. Das ^{238}Pu kann im Gemisch mit ^{239}Pu zur Herstellung von Brennelementen verwendet werden.

Kanada. Die Atomic Energy of Canada Ltd. (AECL) hat ihren ersten Jahresbericht über das kanadische Nuclear Fuel and Waste Management Program veröffentlicht. Dieses Programm umfaßt Untersuchungen zur sicheren Endlagerung radioaktiver Rückstände aus dem Betrieb der CANDU-Reaktoren. In dem Bericht wird festgestellt, daß gute Aussichten für eine sichere und technisch mögliche Endlagerung in tiefen geologischen Formationen bestehen. Die laufenden Forschungs- und Entwicklungsarbeiten sollen vorerst bis 1983 fortgesetzt werden. Gegenwärtig werden neben Felsuntersuchungen vor allem Arbeiten zur Einschmelzung radioaktiver Rückstände in Borosilikatgläser durchgeführt.

Buchbesprechungen

Stabile Isotope und Radioelemente als Indikatoren der Vorgänge der Erzbildung

N. G. Syromjatnikov, L. A. Trofimova, N. I. Zamjatin, Izdatel'stvo "Nauka" Kazachskoj SSR, Alma-Ata 1978 (in Russisch) 247 S., 41 Abb., 59 Tab., 224 Lit.; Rubel 3,00.

Isotopengeochemische Untersuchungen liefern reiches Material für die Deutung der Entstehung von Erzlagerstätten und damit auch für die wissenschaftliche Vorhersage und die Suche von neuen Lagerstätten in weniger bekannten Gebieten. Insbesondere können über das Alter der Vererzung, über die Herkunft der Erze und über die Bedingungen der Bildung und Umbildung Aussagen getroffen werden.

In dem vorliegenden Buch unternehmen die Verfasser den Versuch, ihre langjährigen Forschungen unter einheitlichen Gesichtspunkten darzulegen. Besonders bemerkenswert ist die gleichzeitige Untersuchung von mehreren Indikatoren an den jeweiligen Lagerstätten. Diese Indikatoren sind: die Schwefel-Isotope, die Blei-Isotope und die Elemente Uranium und Thorium. Da die genannten Indikatoren durchaus unterschiedlich reagieren können, wird so das Verständnis für viele Fragen vertieft.

Nach einer kurzen Einleitung über die Methodik der Untersuchungen und die mögliche Deutung der Befunde (etwa 40 Seiten) wird anschließend eine Darstellung der Verhältnisse an Erzlagerstätten Mittelasiens gegeben (etwa 180 Seiten). Zahlreiche Tabellen und Diagramme erleichtern die Lesbarkeit.

Das Buch wird von Geochemikern, Lagerstättenkundlern und Mineralogen mit Interesse gelesen werden.

K. H. Köhler

Das Emanieren von radioaktiven Erzen und Mineralen

V. L. Šaškin, M. I. Prutkina. Atomizdat, Moskva 1979 (in Russisch) 111 S., 6 Abb., 62 Tab., 66 Lit.; broschiert Rubel 1, 20

Die Gammastrahlung der natürlichen Radioelemente Uranium und Thorium stammt vorwiegend von Folgeprodukten des Radons. Messungen der Gammastrahlung, wie sie bei der Suche und Erkundung von Lagerstätten in großem Umfang ausgeführt werden, können deshalb nur dann hinsichtlich des Uraniums oder Thoriums qualitativ und quantitativ richtig gedeutet werden, wenn das Radon im radioaktiven Gleichgewicht mit seinen Vorgängern steht oder wenn die Störung des Gleichgewichtes (wenigstens in der Größenordnung) bekannt ist. Das Verhalten des Radons ist deshalb für die geophysikalische Erkundung von großem Interesse. Es wird im wesentlichen durch den Vorgang des Emanierens beschrieben, der seinerseits von zahlreichen physikalischen Bedingungen abhängt (Größe der Teilchen, Feuchte des Gesteins oder Bodens, Temperatur, Luftdruck und seine Änderung, Vorhandensein von Kapillaren und Poren im Gestein).

In der vorliegenden Arbeit geben die Verfasser, die schon wiederholt in der Literatur mit Untersuchungen zu diesem Gegenstand aufgetreten sind, einen zusammenfassenden Überblick über den gegenwärtigen Stand der Erkenntnis auf diesem Gebiet. Dabei werten sie sowohl die eigenen experimentellen Untersuchungen als auch die zahlreichen Veröffentlichungen der in- und ausländischen Literatur gründlich aus.

Geophysiker, Geochemiker und Festkörperphysiker finden damit eine umfassende und zuverlässige Informationsquelle.

K. H. Köhler

Dispersionsströme von Uranlagerstätten

G. P. Tafeev (Red.). *Atomizdat, Moskva 1979 (in Russisch) 136 S., 44 Abb., 29 Tab., 198 Lit.; broschiert Rubel 1,40*

Der erste Schritt der Suche von Uranlagerstätten besteht darin, daß man Informationen über die mögliche Uranführung größerer Gebiete gewinnt, um anschließend diese Gebiete mit aufwendigeren Verfahren in Einzelheiten zu untersuchen. Für die Übersichtsaufnahme sind geometrische Verfahren besonders geeignet, was auch in ihrer weltweiten Anwendung zum Ausdruck kommt. Beispielsweise sind 1970/71 in den Entwicklungsländern etwa 180000 km^2 und in den entwickelten Ländern etwa 670000 km^2 mit geochemischen Verfahren aufgenommen worden. Die meisten neuen Lagerstätten sind im Gefolge geochemischer Aufnahmen gefunden worden.

In der vorliegenden Veröffentlichung werden die Gesetzmäßigkeiten der Bildung von hydrochemischen und lithochemischen Dispersionsströmen von Uranlagerstätten beschrieben, wie sie von den Autoren in mehr als 20jähriger wissenschaftlicher Arbeit untersucht wurden. Besonders eingehend werden die Migration des Uraniums in wäßriger Phase und seine Akkumulation in Flußbettablagerungen, in hydromorphen Böden und in Torf darg stellt. Auch das Verhalten typischer Begleitelemente un die Bewertung von Befunden für die Sucharbeiten werden ausführlich beschrieben. Für Aufnahmen in den Maßstäben 1:1000000 bis 1:100000 werden detaillierte Hinweise gegeben.

Das Buch wird von Fachleuten der Suche und Erkundung von Erzlagerstätten, insbesondere Uraniumlagerstätten, sicher aufmerksam ausgewertet werden.

K. H. Köhler

H. Johansen, A. Graesser

Model Calculations Related to Separation of Isotopes through Multiphoton Photodissociation of Molecules

Isotopenpraxis **16** (1980) No. 10, p. 330 to 335
6 Fig., – Tab., 7 Ref.

For collision-free SF_6 molecules multiphoton photofragmentation probabilities in intense IR-LASER fields are calculated together with the isotope selectivity of such processes in dependence of the LASER line adopted, the power density, the pulse profile and the temperature. By means of a quantum-mechanical model the rotation-vibration levels of the SF_6 molecule, the detuning degree between LASER frequency and the oscillation frequency of the molecules as well as the intramolecular oscillation relaxation are taken into account.

F. Schneider, H. Unger

On the Determination of Removable Surface Contamination

Isotopenpraxis **16** (1980) No. 10, p. 335 to 339
3 Fig., 6 Tab., 6 Ref.

The determination of removable contamination of materials used in nuclear plants and radionuclide labs is discussed. Above all the problem of the decontamination coefficient achieved by means of the rubbing test is dealt with. The value of the decontamination coefficient depends on the state of the contaminated surface, the state of the contaminant and the removal technique. In practics, insufficient knowledge on the impact of exterior factors frequently prevents exact assessment of the decontamination coefficient. On the basis of presented results, the possibility to introduce a standard decontamination coefficient of 0.5 without affecting the previously achieved quality of radiological protection is outlined.

Letters to the Editor

M. Baer, J. Leonhardt, G. Flachowsky, A. Hennig, I. Wolf, K. Nehring

On High-Energy Irradiation of Straw. Part. 1.

Isotopenpraxis **16** (1980) No. 10, p. 339 to 340
1 Fig., – Tab., 4 Ref.

Information

Isotopenpraxis 16 (1980) No. 10, p. 340 to 343

New Books

Isotopenpraxis 16 (1980) No. 10, p. 343 to 344

Х. Иогансен, А. Грессер

Модельные расчеты по разделению изотопов методом мультифотонного фотодиссоциации молекул

Isotopenpraxis **16** (1980) вып. 10, стр. 330 до 335
6 рис., – табл., 7 лит.

Применительно к свободным от столкновений молекул SF_6 расчетом определяются вероятности мультифотонного фоторасщепления в интенсивном поле ЛАЗЕРА включая изотопной селективности таких процессов в зависимости от принимаемой линии ЛАЗЕРА, плотности мощности, профиля импульсов и температуры. С помощью квантовомеханической модели принимаются в учет вращательно-колебательные уровни молекул SF_6, рассогласование между частотой ЛАЗЕРА и частотой колебаний молекул, а также внутримолекулярное затухание колебаний.

Ф. Шнайдер, Х. Унгер

Об определении снимаемого загрязнения поверхностей

Isotopenpraxis **16** (1980) вып. 10, стр. 335 до 339
3 рис., 6 табл., 6 лит.

Обсуждается вопрос определения снимаемого загрязнений материалов, находящих применение в ядерных установках и радионуклидных лабораториях. Прежде всего освещается проблема коэффициента дезактивации методом обтерения. Величина коэффициента дезактивации зависит от состояния загрязненной поверхности, от состояния загряаняющего вещества и от технологии дезактивации. В практике недостающее знание о воздействии внешних факторов на коэффициент дезактивации часто мешает точной оценке последнего. Поэтому, на основании изложенных результатов указывается на общий коэффициент дезактивации 0,5, который можно внедрить без ущерба для уже достигшего качества радиологической зашиты.

Письма в редакцию

М. Бэр, И. Леонхардт, Г. Флаховский, А. Хенниг, И. Вольф, К. Неринг

О высокоэнергетическом облучении соломы. 1-ое сообщение

Isotopenpraxis **16** (1980) вып. 10, стр. 339 до 340
1 рис., – табл., 4 лит.

Новости

Isotopenpraxis **16** (1980) вып. 10, стр. 340 до 343

Новая литература

Isotopenpraxis **16** (1980) вып. 10, стр. 343 до 344

Contents of this Issue

FSC
www.fsc.org
MIX
Papier | Fördert
gute Waldnutzung
FSC® C083411

Zeitfracht Medien GmbH
Ferdinand-Jühlke-Straße 7
99095 Erfurt, Deutschland
produktsicherheit@kolibri360.de